Image Fusion

H.B. Mitchell

Image Fusion

Theories, Techniques and Applications

H.B. Mitchell
Rehov Brosh 16/2
Mazkaret Batya
Israel
E-mail: harveymitchell@walla.co.il

ISBN 978-3-642-11215-7 e-ISBN 978-3-642-11216-4

DOI 10.1007/978-3-642-11216-4

Library of Congress Control Number: 2009941080

Typesetting: Data supplied by the authors

Production: Scientific Publishing Services Pvt. Ltd., Chennai, India

Cover Design: WMX Design, Heidelberg, Germany

Printed in acid-free paper

9 8 7 6 5 4 3 2 1

springer.com

This book is dedicated to my wife

MARGOT HELENA MITCHELL

A woman of worth who can find? For her price is far above rubies.

Proverbs

Preface

The purpose of this book is to provide a practical introduction to the theories, techniques and applications of image fusion. The present work has been designed as a textbook for a one-semester final-year undergraduate, or first-year graduate, course in image fusion. It should also be useful to practising engineers who wish to learn the concepts of image fusion and apply them to practical applications. In addition, the book may also be used as a supplementary text for a graduate course on topics in advanced image processing.

The book complements the author's previous work on multi-sensor data fusion[1] by concentrating exclusively on the theories, techniques and applications of image fusion. The book is intended to be self-contained in so far as the subject of image fusion is concerned, although some prior exposure to the field of computer vision and image processing may be helpful to the reader.

Apart from two preliminary chapters, the book is divided into three parts. Part I deals with the conceptual theories and ideas which underlie image fusion. Here we emphasize the concept of a common representational framework and include detailed discussions on image registration, radiometric calibration and semantic equalization. Part II deals with a wide range of techniques and algorithms which are in common use in image fusion. Among the topics considered are: sub-space transformations, multi-resolution analysis, wavelets, ensemble learning, bagging, boosting, color spaces, image thresholding, Markov random fields, image similarity measures and the expectation-maximization algorithm. Together Parts I and II provide the reader with an integrated and comprehensive overview of image fusion. Part III deals with applications. In it we examine several real-life image fusion applications. The aim is to illustrate how the theories and techniques of image fusion are used in practical situations.

---

[1] Multi-Sensor Data Fusion: An Introduction by H.B. Mitchell. Published by Springer-Verlag (2007)

As with any other branch of engineering, image fusion is a pragmatic activity which is driven by practicalities. It is therefore important that the reader is able to experiment with the different techniques presented in the book. For this purpose software written in Matlab is particularly convenient. We have therefore included examples of matlab code in the book and also give details of third-party matlab code which may be downloaded from the world wide web. Teachers and students will find additional resources on the author's website: http://www.ee.bgu.ac.il/~harveym.

The book is based on seminars and lectures on image fusion which have been given over the past few years. In particular, the structure and content of the book is based on a graduate course in image fusion which the author has taught at Ben-Gurion University of the Negev. For the opportunity of teaching this course I am particularly indebted to Prof. Mayer Aladjem and to Prof. Dan Sadot. I am also indebted to my wife and children for the support and patience they have shown me while the book was being written.

June 2008
Mazkaret Batya H.B. Mitchell

Contents

Part I: Theories

Chapter 1
Introduction

Abstract. The subject of this book is *image fusion* which we define as the process of combining multiple input images into a single composite image. Our aim is to create from the collection of input images a single output image which contains a better description of the scene than the one provided by any of the individual input images. The output image should therefore be more useful for human visual perception or for machine perception. The basic problem of image fusion is one of determining the best procedure for combining the multiple input images. The view adopted in this book is that combining multiple images with *a priori* information is best handled within a statistical framework. In particular we shall restrict ourselves to classical and robust statistical approaches, Bayesian methods, sub-space and wavelet techniques.

1.1 Synergy

The principal motivation for image fusion is to improve the quality of the information contained in the output image in a process known as *synergy*. A study of existing image fusion techniques and applications shows that image fusion can provide us with an output image with an improved quality. In this case, the benefits of image fusion include:

1. Extended range of operation.
2. Extended spatial and temporal coverage.
3. Reduced uncertainty.
4. Increased reliability.
5. Robust system performance.
6. Compact representation of information.

Traditionally, the input images are captured by the same camera at different times or are captured by different cameras at the same time. However, in the definition of image fusion we shall also include the case when the input images are derived from the same "base" image but which have undergone different processing algorithms.

The following examples illustrate the image fusion synergy process. The first example deals with input images which are captured by the same camera at different times. The next example deals with input images captured by different cameras at the same time, while the third example deals with input images which are derived from the same base image and which are processed differently.

Example 1.1. Multiple Camera Surveillance Systems [9]. The increasing demand for security by society has led to a growing need for surveillance activities in many environments. For example, the surveillance of a wide-area urban site may be provided by periodically scanning the area with a single narrow field-of-view camera. The temporal coverage is, however, limited by the time required for the camera to execute one scan. By using multiple cameras we reduce the mean time between scans and thereby increase the temporal coverage.

Example 1.2. Multispectral Bilateral Video Fusion [2]. A significant problem in night vision imagery is that while an infra-red (IR) image provides a bright and relatively low-noise view of a dark environment, it can be difficult to interpret due to inconsistences with the corresponding visible-spectrum image. In bilateral fusion we enhance a visible video input using information from a spatially and temporally registered IR video input. Our goal is to create a video that appears as if it was imaged only in the visible spectrum and under more ideal exposure conditions than actually existed.

Example 1.3. Color Image Segmentation [11]. A significant problem in computer vision is the reliable segmentation a base image into meaningful labeled segments. In ensemble image segmentation we generate an ensemble of color input images by transforming the base image in different ways. Each input image is separately segmented using a simple segmentation algorithm. By fusing the multiple segmented images we are able to substantially improve both the accuracy and the reliability of the segmentation process.

1.2 Image Fusion Process

Fig. 1.1 shows the principal processes in a generic image fusion processing chain for the case when the output is a single fused image $\tilde{I}$. The principal processes in the chain are:

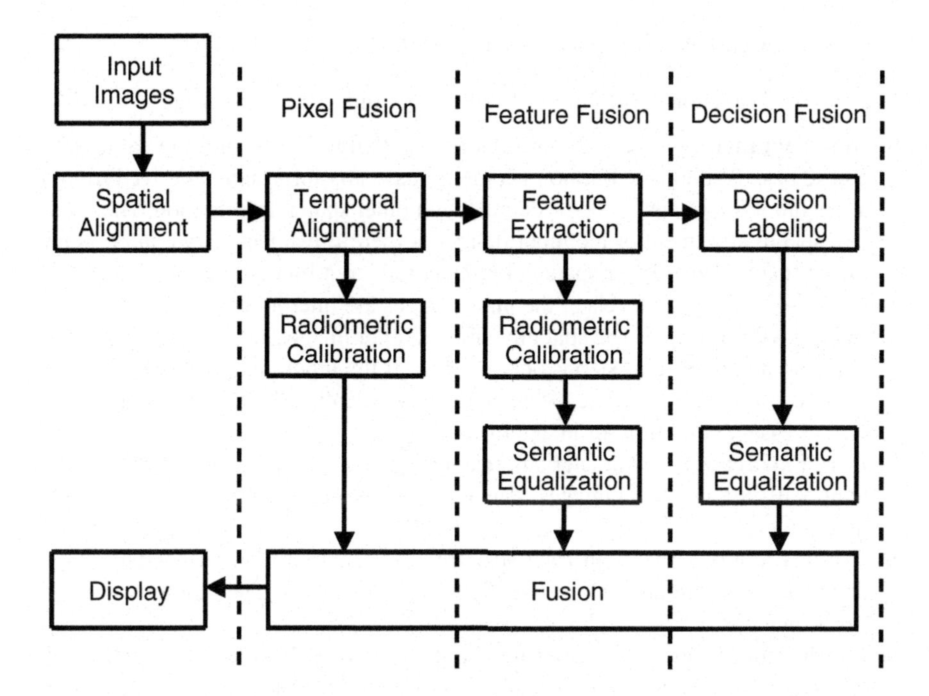

Fig. 1.1 Shows the generic image fusion processing chain. It consists of four principal blocks: (1) Multiple Input Images. Multiple images of the external scene are captured by multiple sensors. (2) Common Representational Format. The input images are transformed into a common representational format. This involves several processes including: spatial and temporal alignment, semantic equivalence, radiometric calibration, feature extraction and decision labeling. (3) Fusion. The multiple images in the common representational format are fused together. The fusion process may be classified into three classes: pixel fusion, feature fusion and decision fusion. (4) Display. The fused output is processed for display.

Multiple Input Images. The external environment is captured by one or more image sensors or cameras. Each camera generates one or more input images.

Common Representational Format. The input images are transformed so they "speak a common language". This involves several processing including: spatial, temporal, semantic and radiometric alignment, feature extraction and decision labeling.

Fusion. After conversion into a common representational format the spatially, temporally, semantically and radiometrically aligned images, feature maps or decision maps are fused together in the fusion block. The output is a fused image $\widetilde{I}$, feature map $\widetilde{F}$ or decision map $\widetilde{D}$.

Display. The fused image, feature map or decision map is processed for display.

1.3 Common Representational Block

The principal function in the common representational format block are:

Spatial Alignment. The input images are spatially aligned into the same geometric base. Without a common geometric base any information derived from a given input image cannot be associated with other spatial information. The accurate spatial alignment of the input images is therefore a necessary condition for image fusion. *Note*. After spatial alignment the input images are re-sampled and if necessary the gray-levels of the input images are interpolated.

Temporal Alignment. The spatially aligned input images are temporally aligned to a common time. This step is only required if the input images are changing or evolving in time. In this case the accurate temporal alignment of the input images is a necessary condition for image fusion.

Feature Extraction. Characteristic features are extracted from the spatially and temporally aligned input images. The output is one or more feature maps for each input image [1].

Decision labeling. Pixels in each spatially and temporally aligned input image or feature map are labeled according to a given criteria. The output is a set of decision maps.

Semantic Equivalence. In order for the input images, feature maps or decision maps to be fused together they must refer to the same object or phenomena. The process of causally linking the different inputs to a common object or phenomena is known as semantic equivalence.

Radiometric Calibration. The spatially, temporally and semantically aligned input images and feature maps are converted to a common measurement scale. This process is known as radiometric calibration.

When the input is a set of K multiple image sequences $I_k(t), t \in [T_1, T_2], k \in \{1, 2, \ldots, K\}$, the output is a fused image sequence $\widetilde{I}(t), t \in [T_1, T_2]$, feature map sequence $\widetilde{F}(t)$ or decision map sequence $\widetilde{D}(t)$. In this case we replace the spatial alignment and temporal alignment blocks in Fig. 1.1 with a single spatial-temporal alignment block which performs both functions simultaneously.

Figs. 1.2–1.4 shows the adaption of the generic processing chain to Ex 1.1–1.3.

1.4 Image Fusion Block

In the image fusion block we fuse together the information contained in the multiple input images after conversion into a common representational format. The common representational format may take the form of an image I, a feature map F or a decision map D. Very often we shall not differentiate between I, F and D and in this case we shall refer to all three as an "image".

We find it convenient to further divide the fusion algorithms into arithmetic, subspace and multi-scale techniques [5].

[1] A feature is any distinguishing property or attribute of an image. Examples of features used in image fusion are: edges, lines, patterns and color.

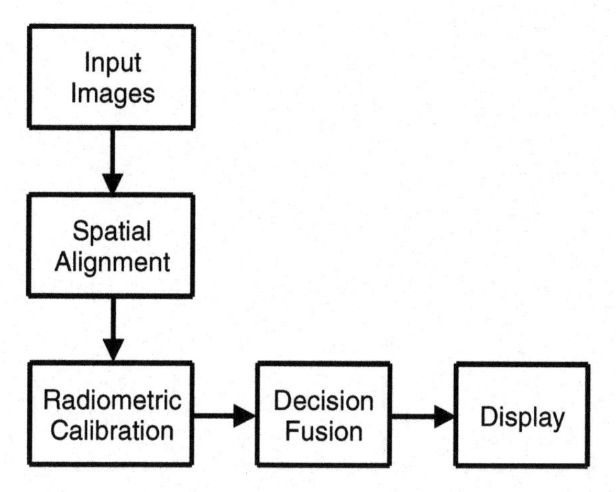

Fig. 1.2 Shows the image processing chain for the multiple camera surveillance system discussed in Ex. 1.1

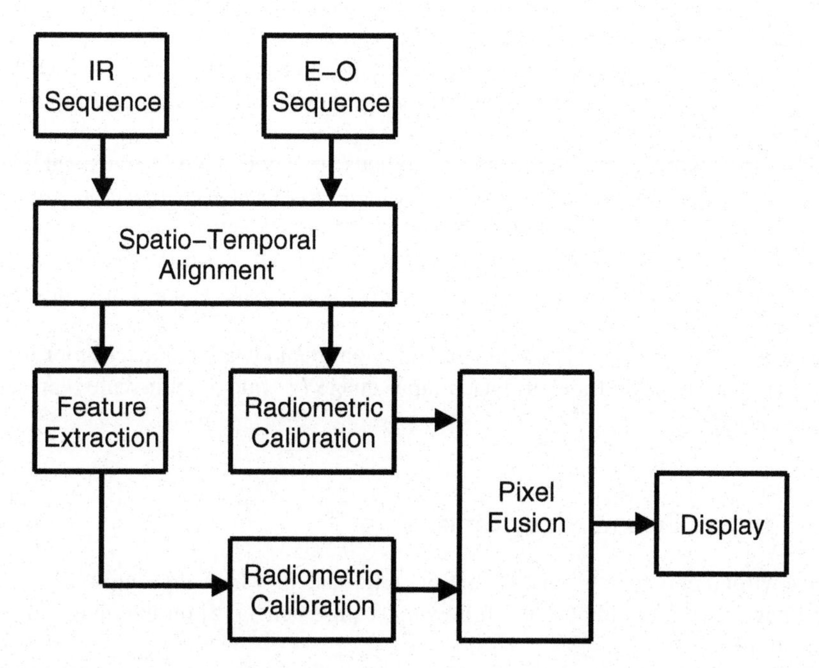

Fig. 1.3 Shows the image processing chain for the multispectral bilateral video fusion system discussed in Ex. 1.2

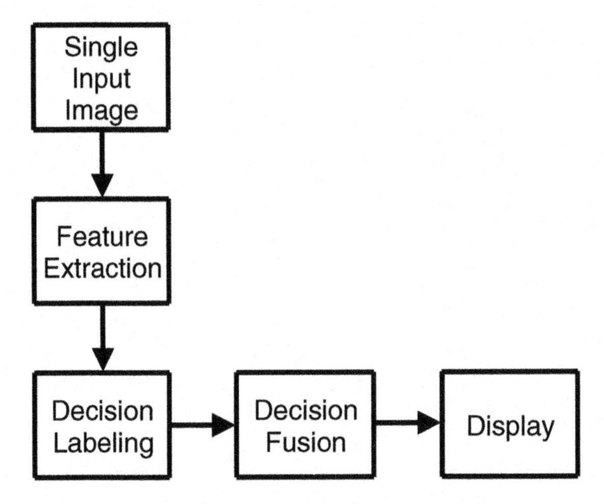

Fig. 1.4 Shows the image processing chain for the color image segmentation system discussed in Ex. 1.3

Pixel Operations. These operations include simple arithmetic operators such as addition, subtraction, division and multiplication as well as minimum, maximum, median and rank. It also includes more complicated operators which are defined by a function or algorithm, such as the expectation-maximization algorithm and Markov random field.

Sub-Space Methods. The sub-space methods are a collection of statistical techniques which remove the correlation which exists between the input images $I_k, k \in \{1,2,\ldots,K\}$. Important sub-space techniques are: principal component analysis (PCA), independent component analysis (ICA), non-negative matrix factorization (NMF), canonical correlation analysis (CCA) and linear discriminant analysis (LDA).

Multi-Scale Methods. The multi-scale fusion methods are a collection of techniques in which we transform each input image $I^{(k)}$ into a multi-scale representation: $\left(y_0^{(k)}, y_1^{(k)}, \ldots, y_L^{(k)}\right)$.

1.5 Image Fusion Algorithms

For environments which are essentially static and in which the output is a single image $\tilde{I}$ we often impose the following requirements [15] on the image fusion algorithms:

Pattern conservation. The fusion process should preserve all relevant information on the input imagery in the composite image.

Artifact free. The fusion scheme should not introduce any artifacts or inconsistencies which would distract the human observer or subsequent image processing stages.

Invariance. The fusion scheme should be shift and rotational invariant, i.e. the fusion result should not depend on the location or orientation of an object in the input imagery.

For environments which are evolving in time the input is a set of input sequences $I_k(t), t \in [T_1, T_2], k \in \{1, 2, \ldots, K\}$ and the output is a fused image sequence $\widetilde{I}(t)$. In this case, we often impose the following additional requirements on the image fusion algorithms:

Temporal stability. The fused output should be temporally stable, that is, gray-level changes in $\widetilde{I}(t)$, should be present in at least one of the input sequences $I_k(t)$.

Temporal consistency. Gray level changes which occur in the input sequences $I_k(t)$ must be present in the fused sequence $\widetilde{I}(t)$.

1.6 Organization

Apart from two preliminary chapters, the book is divided into three parts:

Part I: Theories. This consists of Chapts. 3-7 and deals with the conceptual theories and ideas which underlie image fusion. Here we emphasize the concept of a common representational framework and include detailed discussions on image registration, radiometric calibration and semantic equalization.

Part II: Techniques. This consists of Chapts. 8-18 and deals with a wide range of techniques and algorithms which are in common use in image fusion. Among the topics considered are: sub-space transformations, multi-resolution analysis, ensemble learning, bagging, boosting, color spaces, Markov random fields, image similarity measures and the expectation-maximization algorithm. Together Parts I and II provide the reader with an integrated and comprehensive overview of image fusion.

Part III: Applications. This consists of Chapts. 19-22 and deals with applications. In it we examine several real-life image fusion applications. The aim is to illustrate how the theories and techniques of image fusion are used in practical situations.

1.7 Software

The following matlab routines and toolboxes are of general utility and are widely used in image fusion.

IMAGE FUSION TOOLKIT. Image fusion toolbox. Author: Eduardo Fernandez Canga.

MATIFUS. Matlab toolbox for image fusion. Authors: P. M. de Zeeuw, G. Piella and H. J. A. M. Heijmans [14].

MATLAB IMAGE PROCESSING TOOLBOX. Matalb image processing toolbox.

MATLAB WAVELET TOOLBOX. Matlab wavelet toolbox.

1.8 Further Reading

General overviews on multi-sensor image fusion are [1, 3, 4, 5, 6, 7, 8, 10, 12]. For an extended discussion regarding the issues involved in defining multi-sensor image fusion and related terms, see [13, 16].

References

1. Appriou, A., Ayoun, A., Benferhat, S., Bernard, P., Cholvy, L., Cooke, R., Cuppens, F., Dubois, D., Fargier, H., Grabisch, M., Kruse, R., Lang, J., Moral, S., Prade, H., Saffiotti, A., Smets, P., Sossai, C.: Fusion: general concepts and characteristics. Int. J. Intell. Syst. 16, 1107–1134 (2001)
2. Bennett, E.P., Mason, J.L., McMillan, L.: Multispectral bilateral video fusion. IEEE Trans. Image Process. 16, 1185–1194 (2007)
3. Bloch, I. (ed.): Information Fusion in Signal and Image Processing. Wiley, Chichester (2008)
4. Fenoy, G.P.: Adaptive wavelets and their applications to image fusion and compression. PhD thesis University of Amsterdam (2003)
5. Gang, H.: Image fusion, image registration and radiometric normalization for high resolution image processing. PhD thesis, University of New Brunswick, Canada (2007)
6. Hall, D.L., Llinas, J. (eds.): Handbook of Multisensor Data Fusion. CRC Press, Boca Raton (2001)
7. Hall, D.L., McMullen, S.: Mathematical Techniques in Multisensor Data Fusion, 2nd edn. Artech House Inc., Norwood (2004)
8. Hyder, A.K., Shahbazian, E., Waltz, E. (eds.): Multisensor Fusion. Kluwer Academic Publishers, Dordrecht (2002)
9. Jones, G.D., Allsop, R.E., Gilby, J.H.: Bayesian analysis for fusion of data from disparate imaging systems for surveillance. Image Vis. Comp. 21, 843–849 (2003)
10. Luo, R.C., Yih, C.-C., Su, K.L.: Multisensor fusion and integration: approaches, applications and future research directions. IEEE Sensors Journal 2, 107–119 (2002)
11. Mignotte, M.: Segmentation by fusion of histogram-based K -means clusters in different clolor spaces. IEEE Trans. Image Process. 17, 780–787 (2008)
12. Mitchell, H.B.: Multi-Sensor Data Fusion: An Introduction. Springer, Berlin (2007)
13. Oxley, M.E., Thorsen, S.N.: Fusion or integration: what's the difference? In: Proc. 7th Int. Conf. Information Fusion, pp. 429–434 (2004)
14. Piella, G.: A general framework for multiresolution image fusion: from pixels to regions. Inform. Fusion 9, 259–280 (2003)
15. Rockinger, O., Fechner, T.: Pixel-level image fusion. In: Proc. SPIE, vol. 3374, pp. 378–388 (1998)
16. Wald, L.: Some terms of reference in data fusion. IEEE Trans. Geosci. Remote Sensing 37, 1190–1193 (1999)

Chapter 2
Image Sensors

Abstract. The subject of this chapter is the image sensor or camera. This is a special device which interacts *directly* with the environment and is ultimately the source of all the input data in an image fusion system [3]. The image sensor may be any device which is capable of perceiving a physical property, or environmental attribute, such as heat, light, sound, pressure, magnetism or motion. However, to be useful, the sensor must map the value of the property or attribute to a quantitative measurement in a *consistent* and *predictable* manner. To make our discussion more concrete we shall limit ourselves to digital image sensors or cameras.

2.1 Digital Camera

A digital camera records the time-and space-varying light intensity information reflected and emitted from objects in a three-dimensional physical scene. The camera consists of two parts: (1) An optical system whose purpose is to form a two-dimensional image of the electromagnetic radiation emitted and reflected from objects in the three-dimensional scene. (2) A recording system whose purpose is to measure and record the characteristics of the radiation incident at the focal plane. Furthermore, the incident electromagnetic radiation typically consists of a range of wavelengths, so the incident energy may be measured in one or more spectral wavelength ranges or bands.

2.2 Optical System

The optical system typically consists of a series of lenses which serve to focus the illumination on a two-dimensional surface, called the focal plane, where the characteristics of the incident radiation may be recorded. Though the image formation properties of the optical system are typically complicated by the presence of various distortions or aberrations, it is useful to model it using an idealized geometric model which projects from locations in the three-dimensional world $R(t) = (X(t),Y(t),Z(t))^T$ to two-dimensional locations $r(t) = (x(t),y(t))^T$ in the focal plane.

The most commonly used models of the image projection characteristics are the perspective projection and the simpler orthographic projection.

2.2.1 Perspective Projection

Perspective projection models the geometric projection characteristics of an idealized pinhole camera as shown in Fig. 2.1. If we assume the origin of the 3D scene

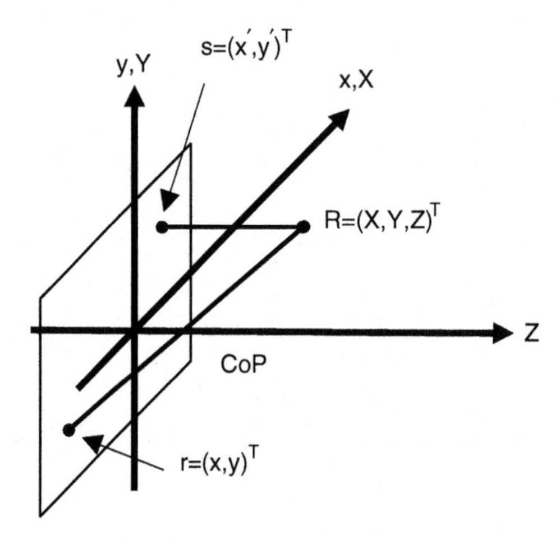

Fig. 2.1 Shows a point $R = (X,Y,Z)^T$ in three-dimensional space and its perspective projection onto the focal plane at $r = (x,y)^T$ and its orthographic projection on to the focal plane at $s = (x',y')$.

and the 2D image plane coincide, then the perspective projection of $R = (X,Y,Z)$ is $r = (x,y)$, where $x = fX/(f-Z)$, $y = fY/(f-Z)$ and f is the focal length, which is the distance along the optical axis from the image plane to the center of projection (CoP).

2.2.2 Orthographic Projection

Orthographic projection assumes a parallel projection of the 3D scene onto the image plane as shown in Fig. 2.1. If we assume the origin of the 3D scene and the 2D image plane coincide, then the orthographic projection of $R = (X,Y,Z)$ is $s = (x',y')$, where $x' = X$ and $y' = Y$.

2.3 Recording Systems

In a digital camera the time-varying focal plane image is recorded using a focal plane of light-sensitive elements or pixels. The spectral response of the sensor elements is determined by the material characteristics. Color reproduction is typically achieved with the use of a color filter array (CFA) which ensures that individual pixels measure the wavelengths in a specified spectral band.

2.3.1 Noise

The recorded image intensity information is always affected by noise in the detection and recording system. In imaging systems utilizing focal plane sensor arrays, the electronic properties of these devices are a source of noise. In situations of extremely low illumination, or in certain medical imaging applications, the number of incident photons is so small that a Poisson model for the photon count is necessary. This results in what is referred to as photon counting noise. Other sources of noise include quantization, speckle and atmospheric effects.

Example 2.1. Noise Models [1]. Noise estimation is an important task in many image fusion applications. It is common practice to model the noise in an image as either additive or multiplicative. The corresponding noise models are:

$$I(i,j) = \begin{cases} I_0(i,j) + n(i,j) & \text{additive}\,, \\ I_0 n(i,j) & \text{multiplicative}\,, \end{cases}$$

where $I_0(i,j)$ is the gray-level at pixel (i,j) in the noise-free image and $I(i,j)$ is the gray-level at pixel (i,j) in the noisy image.

2.4 Sampling

The electromagnetic radiation incident at the focal plane is a function of four continuous variables: two spatial variables x and y, a temporal variable t, and wavelength λ. At the focal plane the incident radiation is sampled[1] in three domains:

Spatial sampling. The spatial variation of light intensity is recorded at a finite set of locations. In the focal plane arrays the sampling density is determined by the number of discrete locations or pixels where photons are collected.

---

[1] In real systems, sampling the focal plane radiation involves integration of the value of the function in a neighborhood surrounding the sampling location. For spatial sampling, this implies integration of the function over the spatial variable(s). In a CCD focal plane array sensor for example, each pixel accumulates the charge generated by photons which strike the light-sensitive area of the pixel. For functions of time, the integration is over the temporal variable. Integration over wavelength is usually a side effect of the fact that sensing devices and materials respond to photons in a range of wavelengths rather than at discrete wavelengths.

Temporal sampling. The variation of the image as a function of time is recorded. This is achieved by sampling the focal plane image at regularly spaced time instants.

Wavelength sampling. The radiation incident at the focal plane is recorded in different wavelength bands. For visual images we use three wavelength bands which correspond to the human visual system's perception of the colors red, green and blue.

2.4.1 *Quantization*

In a digital camera the continuous-valued intensity $f(x,y)$ is quantized to values drawn from a finite set of L reconstruction levels l:

$$Q(f(x,y)) = l \, . \tag{2.1}$$

Since the set of reconstruction levels is finite, the reconstruction levels may be put in correspondence with the natural numbers $l \in \{0,1,\ldots,L-1\}$. This makes the representation amenable to finite word-length representation and processing by digital computer [2].

2.4.2 *Bayer*

For color images in which we have three spectral bands corresponding to R, G and B, we often use reduced spatial sampling rates. One example is the Bayer color filter array (CFA) in which the R, G and B bands are sub-sampled and formed into a single Bayer image as shown in Fig. 2.2

R	G	R	G	R	G	R
G	B	G	B	G	B	G
R	G	R	G	R	G	R
G	B	G	B	G	B	G

Fig. 2.2 Shows the arrangement of the R, G and B filters in the Bayer color filter array

[2] For 8-bit gray-scale input image, the reconstruction levels are $\{0,1,\ldots,255\}$. For a 24-bit RGB image, the reconstruction levels of each color plane are $\{0,1,\ldots,255\}$.

Example 2.2. CCD Cameras [6]. A typical CCD camera imaging process is shown in Fig. 2.3. Under normal weather conditions the camera irradiance, i. e. the energy incident on the CCD is proportional to the attenuated scene radiance. The lens system transmits this camera irradiance unevenly because of several factors which are collectively known as *vignetting*. The light energy transmitted through the lens system is then converted to electrons in the CCD array. In color CCD cameras, a color filter array (CFA) is often used in the sensor. To obtain a full-resolution color image, we perform some form of interpolation on the CFA image. This operation is followed by white balancing, a process that attempts to generate output images that are consistent with human perception of color under different illumination conditions. If we model the white balancing as a separate transformation on each channel involving a scaling of values and an offset. Then the output of the CCD unit is often modeled as

$$y = aE + N_s + N_c + b\ .$$

where E is the ideal CCD unit response, a is a combined scaling factor that incorporates the scale due to white balancing and the exposure time Δt, N_s and N_c are shot and thermal noise.

To account for the non-linear transfer function relating the input voltage in the display monitor to the display brightness, γ correction is usually applied to the sensed image value. Including other noise effects such as quantization error, amplifier noise, D/A and A/D noise, as an additional noise term N_q with a constant variance, we have the expression for the observed image measurements:

$$z = f(aE + N_s + N_c + b) + N_q\ .$$

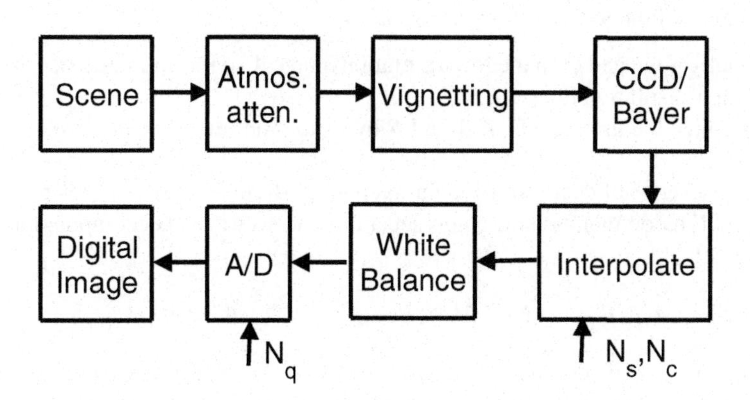

Fig. 2.3 Shows a typical CCD camera imaging pipeline

2.5 Spatial vs. Spectral Resolution

The demand for high resolution images is increasing all the time. The high-resolution images provide the viewer with more details which are often critical. The simplest way to obtain a high-resolution image is to reduce the pixel size and thereby increase the number of pixels per unit area. However, since a reduction of pixel size causes a decrease in the amount of light, shot noise is generated that severely degrades the image quality. As a result many imaging systems have been designed that allow for some level of aliasing during image aquisition.

2.5.1 Spatial Resolution

The spatial resolution of an analog imaging system is characterized by the point spread function (PSF) of the whole observing system. This includes not only the optics and the detector system but also atmospheric conditions and the illumination of the scene. In digital image sensors, the analog images produced by the optical system are spatially sampled by the detector. If the images are oversampled with a frequency higher than the Nyquist frequency, then the resolution properties are preserved. However, in practice, most digital image sensors undersample the analog signal. As a consequence the resulting resolution is determined by the spatial sampling frequency and the PSF. In satellite-based cameras, the resolution is fixed mainly by the sampling frequency and is thus directly related to the area in ground that represents a pixel in the detector [5].

The following example illustrates a simple method for estimating the relative spatial resolution between two input images.

Example 2.3. Estimating the Relative Resolution Between Two Input Images [4]. Let A and B denote two input images of the same area, where A has the low spatial resolution and B has the high spatial resolution. We estimate the relative resolution between A and B as follows.

1. Spatially align and crop the low-resolution image A so it is the same size as the high-resolution image, B.
2. Normalize the grey levels of A and B by histogram matching the two images.
3. Create a series of decreasing resolution images $B_1, B_2, \ldots, B_L$. For this purpose, [4] recommends using a non-separable *a trous* wavelet decomposition scheme:

$$B_1 = M_1 \otimes B\,, \qquad B_2 = M_2 \otimes B_1\,, \qquad \ldots \qquad B_L = M_L \otimes B_{L-1}\,.$$

where $M_{l+1} \otimes B_l$ denotes the convolution of B_l with a mask M_{l+1} (see Chapt. 8). Often we use dyadic masks. In this case, the input image B has double the resolution of B_1, the image B_1 has double the resolution of B_2, and so on.

4. Measure the correlation ρ_l between A and $B_l, l \in \{1,2,\ldots,L\}$.
5. Let $l^* = \arg\max_l(\rho_l)$ denote the image B^* with the maximum correlation coefficient.
6. From this point of maximum correlation compute the relative resolution between the two input images A and B (Fig. 2.4).

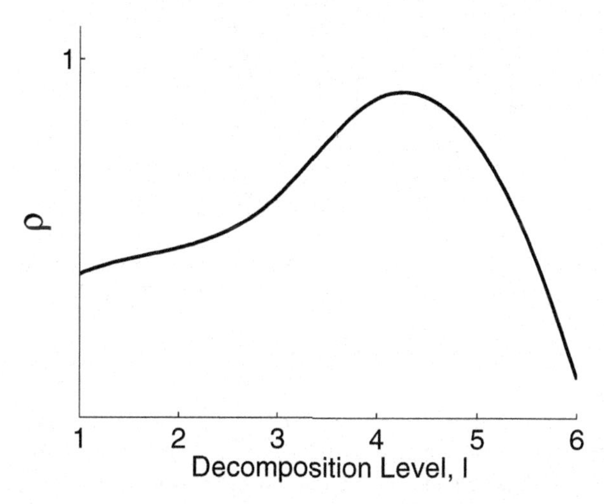

Fig. 2.4 Shows a typical correlation curve as a function of the wavelet decomposition level l

2.5.2 *Spectral Resolution*

Spectral resolution is the width within the electromagnetic spectrum that can be sensed by a band in a sensor. As the spectral bandwidth becomes narrower, so the spectral resolution becomes higher. The spectral resolution is determined by the spectral response function of the sensor element which is defined as the probability that a photon at a given frequency is detected by the sensor. Fig. 2.5 shows a typical spectral response function of the IKONOS sensor.

In fusing together different images of different modalities, it is important to take into account the spectral response function of each image.

Example 2.4. Pan Sharpening. Physical constraints mean that remote sensing images are usually of two distinct types: high spatial resolution panchromatic images and low spatial resolution multi-spectral images. By means of image fusion (pan-sharpening methods) we try to recover the image obtained by an ideal sensor which has the same spectral sensitivity of the multi-spectral sensor and the same spatial resolution of the panchromatic sensor. In order to do this the latest pan-sharpening algorithms take into account the spectral response of the two sensors (see Sect. 19.6).

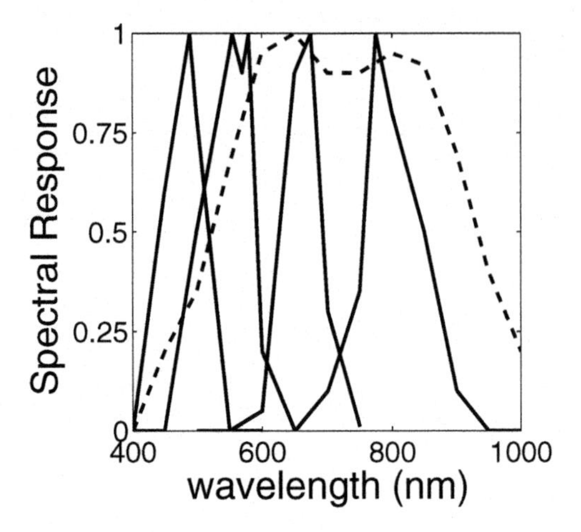

Fig. 2.5 Shows the spectral response function for IKONOS sensor. The full lines show the blue, green, red and near infra-red spectral responses. The dashed line shows the panchromatic spectral response.

We now derive an important relationship between the spectral response functions $\phi_A(\nu)$ and $\phi_B(\nu)$ of two sensors A and B. Let a be the detection of a photon by sensor A and b be the detection of a photon by sensor B. Then the probability of these events are, respectively,

$$P(a) = \int \phi_A(\nu)d\nu \qquad \text{and} \qquad P(b) = \int \phi_B(\nu)d\nu \, .$$

Given a photon is detected by one sensor, then the probability it is detected by the second sensor is

$$P(b|a) = \frac{\int \min\left(\phi_A(\nu), \phi_B(\nu)\right) d\nu}{P(a)} \, ,$$
$$P(a|b) = \frac{\int \min\left(\phi_B(\nu), \phi_A(\nu)\right) d\nu}{P(b)} \, .$$

If n_A and n_B are, respectively, the number of photons detected by the two sensors, then the number of photons simultaneously detected by both sensors is

$$n_{AB} = P(a|b)n_B = P(b|a)n_A \, .$$

Combining these equations we can predict the number of photons that one sensor should detect given the number of photons detected by the other sensor:

$$\hat{n}_A = \frac{P(a|b)}{P(b|a)} n_B = \frac{\int \phi_A(v)dv}{\int \phi_B(v)dv} n_B ,$$
$$\hat{n}_B = \frac{P(b|a)}{P(a|b)} n_A = \frac{\int \phi_B(v)dv}{\int \phi_A(v)dv} n_A . \tag{2.2}$$

2.6 Further Reading

A general reference on the use of image sensors in image fusion is [3]. CCD image sensors are discussed in [6]. A detailed description of image formation for thermal image sensor is given in [2].

References

1. Aja-Fernandez, S., Vegas-Sanchez-Ferrero, G., Martin-Fernandez, M., Alberola-Lopez, C.: Image Vis. Comp. Automatic noise estimation in images using local statistics. Additive and multiplicative cases 27, 756–770 (2009)
2. Bobrov, S., Schechner, Y.Y.: Image-based prediction of imaging and vision performance. J. Opt. Soc. Am. 24A, 1920–1929 (2007)
3. Fowler, K.R., Schmalzel, J.L.: Sensors: the first stage in the measurement chain. IEEE Instrument. and Measure. Mag., 60–65 (September 2004)
4. Nunez, J., Fors, O., Otazu, X., Pala, V., Arbiol, R., Merino, M.T.: A wavelet-based method for the determination of the relative resolution between remotely sensed images. IEEE Trans. Geosci. Remote Sensing 44, 2539–2548 (2006)
5. Thomson, G.H.: A note on spatial resolution measurement and its implications for image radiometry. Int. J. Remote Sensing 30, 1–8 (2009)
6. Tsin, Y., Ramesh, V., Kanade, T.: Statistical calibration of CCD imaging process. Int. J. Comp. Vis. 1, 480–487 (2001)

Part I
Theories

Chapter 3
Common Representational Format

Abstract. The subject of this chapter is the common representational format. Conversion of all sensor observations to a common format is a basic requirement for image fusion. The reason for this is that only after conversion to a common format are the input images compatible, i. e. the input images "speak a common language" and image fusion may be performed. In this chapter we shall consider the principal theories and techniques which underlie the concept of a common representational format.

3.1 Introduction

Conversion of all input images to a common representational format is a basic requirement for image fusion. The reason for this is that only after conversion to a common format are the input images compatible and fusion may take place.

The following example illustrates the concept of a common representational format in brain research.

Example 3.1. A Standardized Brain Atlas: A Common Representational Format for Brain Research [14]. In order to compare different brains and, to facilitate comparisons on a voxel-by-voxel basis, we use a standardized anatomically-based coordinate system or *brain atlas*. The idea is that, in the new coordinate system, all brains have the same orientation and size. The transformation to this coordinate system also gives us the means to enhance weak, or noisy, signals by averaging the transformed images. The standardized brain atlas allows us to catalogue the anatomical, metabolic, electrophysiological, and chemical architecture of different brains into the same coordinate systems.

The process of converting the input images into a common representational format involves many different processes. As listed in Chapt. 1, the principal processes are:

Spatial Alignment. Transformation of the local spatial positions (x,y) to a common coordinate system. The process is often performed using a statistical matching technique and is considered in Chapt. 4.

Temporal Alignment. Transformation of the local times t to a common time axis. The process is often performed using a dynamic time warping algorithm. In many image fusion applications, the spatial and temporal algnments are performed simultaneously in a joint spatial-temporal alignment algorithm.

Feature Extraction. Transformation of the input images into corresponding feature maps. The process is performed by extracting characteristic features from the input images. Different feature extraction processes are considered in Chapts. 8, 9, 12, 13 and 16.

Decision labeling. Transformation of the input image into a corresponding decision, or label, map. The process is performed by applying decision operators on the input images or on the feature maps. The process of decision labeling is considered in Chapts. 9, 12, 16 and 17.

Semantic Equivalence. Transformation of the input images so the pixel values all refer to the same object or phenomena. This is considered in Chapt. 5.

Radiometric Calibration. Transformation of the input images so the pixel values are all measured on a common scale. The process is considered in Chapt. 6.

In many image fusion applications, the construction of a common coordinate system is the primary fusion algorithm. The following example illustrates the construction of a common representational format for an environment which is essentially static and in which the sensors are all of the same type. In this case, temporal alignment, semantic equivalence and radiometric calibration are not required and the construction of a common representational format reduces to the construction of a common spatial coordinate system.

Example 3.2. A Distributed Surveillance System [15]. The demand for surveillance activities for safety and security purposes has received particular attention for remote sensing in transportation applications (such as airports, maritime environments, railways, motorways) and in public places (such as banks, supermarkets, department stores and parking lots). Such systems typically consist of a number of video-based television cameras located in multiple locations. Consider a sequence of M narrow field-of-view "spot" images $I_m, m \in \{1,2,\ldots,M\}$, taken of a wide surveillance area.

We establish a common coordinate system by building a panoramic or "mosaic" image I^* from the sequence of images I_m (Fig. 3.1). For each image I_m, we find a geometric transformation T_m which maps the local "camera-centered" coordinate system of I_m to the common "object-centered" coordinate system of I^*. We then form the mosaic image I^* by "stitching" or "compositing" together the transformed images $T_m(I_m)$. In this case, the aim of a stitching algorithm is to produce a visually plausible mosaic image I^* in

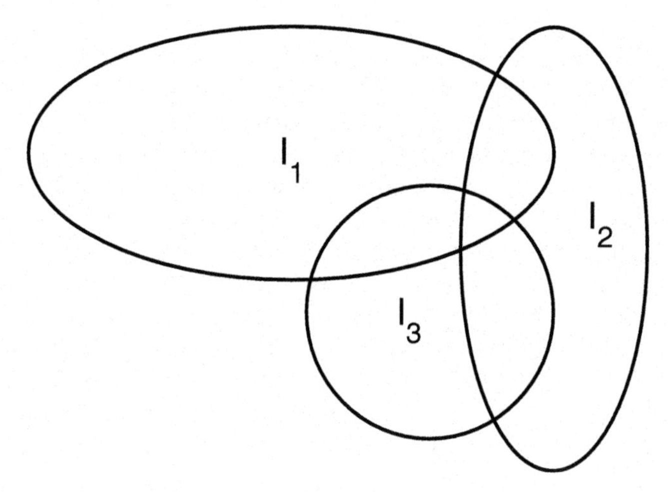

Fig. 3.1 Shows the surveillance of a wide-area site with a sequence of "spot" images I_1, I_2 and I_3. Each spot image I_m is transformed to a common coordinate system using a transformation T_m. The union of three three images define the mosaic image I^*. The stitching algorithm is used in those regions where two, or more, transformed images $T_m(I_m)$ overlap.

which, geometrically and photometrically, I^* is as similar as possible to the input images $T_m(I_m)$ and the seams between the stitched images are invisible.

3.2 Geographical Information System

An important example of a common representational format is a *Geographical Information System.* In a Geographic Information System (GIS) we combine multiple images of the earth obtained from many different sensors and maps, including demographic and infrastructure maps, into a common coordinate system.

3.3 Choosing a Common Representational Format

The common representational format plays a crucial role in image fusion. In fact the choice of common representational format will often govern the fusion algorithm which is used and its performance. In the sections which follow we consider some of the issues involved in choosing an appropriate common representational format.

3.3.1 Human Fusion

In some applications image fusion is performed by a human observer. In this case, the common representational format is chosen as an aid to the human observer.

Example 3.3. Myocardial Imaging [3]. Polar maps, or "bull's-eye" images, are a standard way of displaying myocardial functions and are well established in clinical settings. Fig. 3.2 shows the polar image representation of a left ventricle. The polar maps are constructed by combining images from multiple planes so that information about the entire myocardium can be displayed in a single image. Polar maps can be compared to a three-dimensional cone-shaped heart activity image projected onto a single plane. Each image plane forms a ring in the polar map. Although the rings may be divided into an arbitrary number of sectors, in practice, a clinician uses four (anterior, lateral, inferior and septal) or six (anterior, anterior-lateral, inferior-lateral, inferior, inferior-septal and anterior-septal) sectors for his visual interpretation of the image.

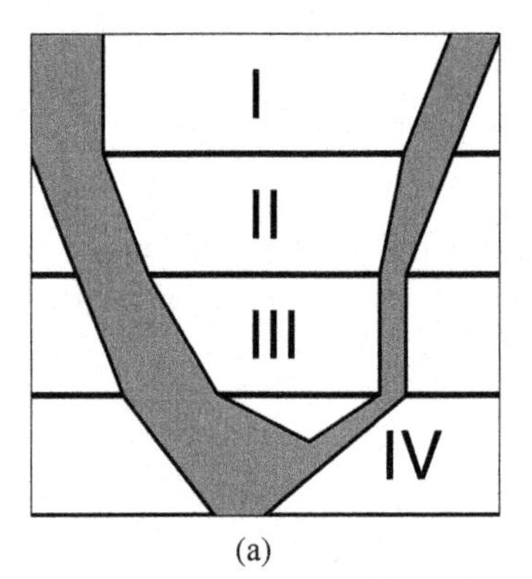

(a)

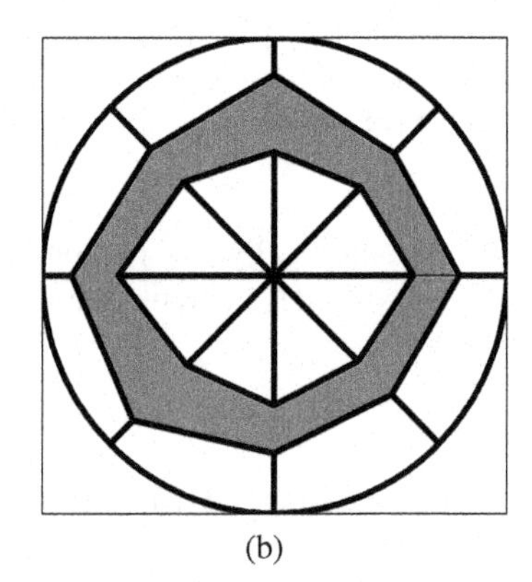

(b)

Fig. 3.2 Shows a polar image of a left ventricle. **(a)** Shows the the left ventricle divided into four slices: I (Basal), II (Mid-Basal), III (Mid-Apical) and IV (Apical). **(b)** Shows a given slice divided into 8 sectors.

3.3.2 Sparseness

In some applications only a small number of image pixels are transformed into the common representational format. The result is a sparse representation. The following example illustrates the conversion of an input image into a sparse representation which is scale and rotation invariant and is invariant across a wide range of distortions and changes in illumination.

Example 3.4. Scale Invariant Feature Transform (SIFT) [9, 12]. SIFT key-points (see Chapt. 13) are invariant to image scale and rotation and provide robust matching across a substantial range of affine distortion changes in three-dimensional viewpoint, occlusion, noise and changes in illumination. In addition, the features are highly distinctive in the sense that a single feature can be reliably matched with only a few other SIFT key-points. The SIFT common representational format is a vector containing $L = 128$ components:

$$\mathbf{s} = (s(1), s(2), \ldots, s(128))^T .$$

It is computed by partitioning the image region surrounding each key-point into a 4×4 grid of sub-regions, and computing an 8-bin orientation histogram in each subregion. The 128-component vector is then formed by concatenating the 16 orientation histograms.

3.3.3 Object Recognition

In object recognition applications we require a common representational format which is invariant to translation, rotation and scaling and is insensitive to variations due to articulation, occlusion and noise (Fig. 3.3). The following example illustrates shape-context. This is a common representational format which is used for comparing two-dimensional shapes and contours.

Example 3.5. Shape Context [4]. Let C denote a closed contour in a given input image. Its shape context S_c is defined as follows. Let $z_i = (x_i, y_i), i \in \{1, 2, \ldots, M\}$, denote a set of M sample points on the contour C (Fig. 3.4). Then $S_c = (S_1, S_2, \ldots, S_M)$, where S_i is a two-dimensional histogram which describes the distribution of the distances and orientations of the points $z_j, j \neq i$, relative to z_i.

Let r_{ij} and θ_{ij} denote the distance and orientation of the point z_j relative to z_i. To define θ_{ij} we require an axis. Two rotationally invariant axes which are commonly used for this purpose are:

(1) The tangent of the contour at z_i,
(2) The line which joins the center of mass of the contour and the point z_i.

In Fig. 3.4 we show the first option, i. e. we measure the angles θ_{ij} relative to the tangent of the contour at z_i.

We divide the (r, θ) space into L vertical columns $\Theta_l, l \in \{1, 2, \ldots, L\}$, and K horizontal rows $R_k, k \in \{1, 2, \ldots, K\}$. If $h(k, l)$ is the number of points $z_j, j \neq i$, for which r_{ij} lies in the kth row and θ_{ij} lies in the lth column, then

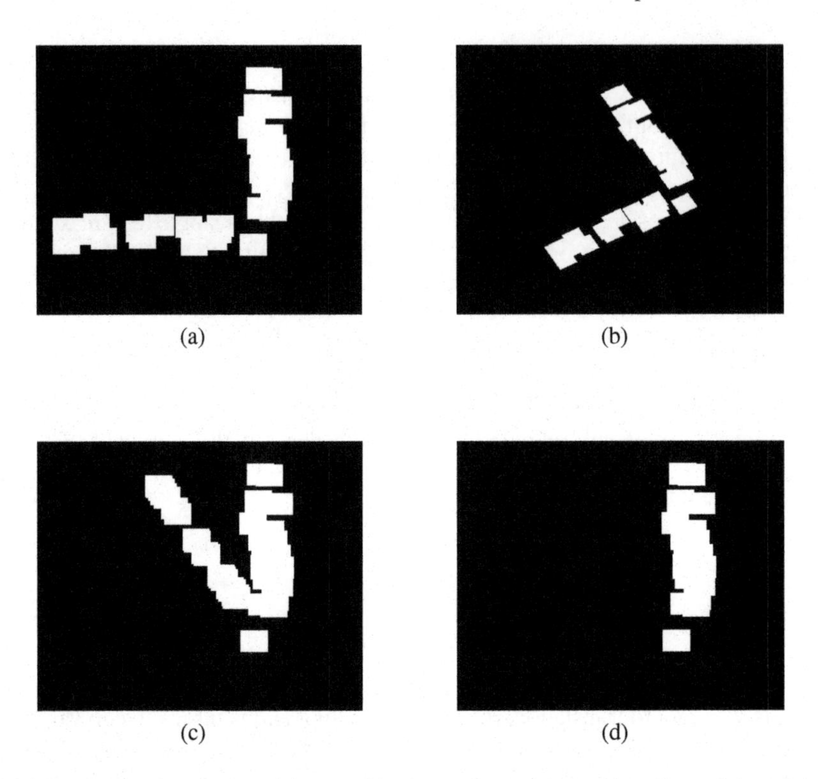

Fig. 3.3 **(a)** Shows the original image. **(b)** Shows the rotation and scaling of the original image.**(c)** Shows the articulation of the original image.**(d)** Shows the occlusion of the original image.

$$S_i = \begin{pmatrix} h(1,1) & h(1,2) & \dots & h(1,K) \\ h(2,1) & h(2,2) & \dots & h(2,K) \\ \vdots & \vdots & \ddots & \vdots \\ h(K,1) & h(K,2) & \dots & h(K,L) \end{pmatrix} .$$

The shape context $S_c = (S_1, S_2, \dots, S_M)$ is invariant to translation and rotation. To a good approximation it is also invariant to occlusion and to noise. Although S_c is not scale invariant it may be made scale invariant by dividing the distances r_{ij} through with the mean distance $\bar{r}$, where

$$\bar{r} = \frac{1}{M(M-1)} \sum_{i=1}^{M} \sum_{j=1, j \neq i}^{M} r_{ij} .$$

Although widely used for comparing two-dimensional shapes and contours (see Ex. 5.6) it is less successful when comparing articulated objects. In this case we may use a modified shape context common representational representation known as the inner distance [8].

Example 3.6. Inner Distance [8]. Suppose the object shown in Fig. 3.4 is in fact an articulated object consisting of two parts A and B which are hinged together at X. Then the inner distance is defined as follows: For any two points z_i and z_j the inner distance is equal to the shape context except we replace r_{ij} with R_{ij}, where R_{ij} is the length of the shortest jointed line which joins z_i and z_j and which lies within the given contour. To a good approximation the inner distance is invariant as we change the angle between A and B.

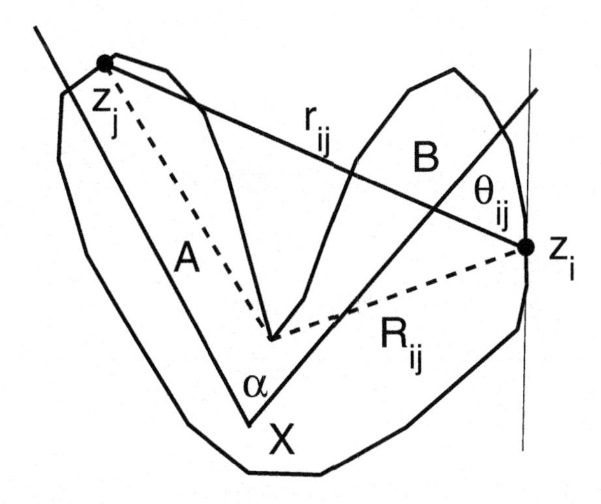

Fig. 3.4 Shows a closed contour defined by M points $z_i, i \in \{1,2,\ldots,M\}$. The line joining the points z_i and z_j has a Euclidean length r_{ij} and it makes an angle θ_{ij} with the tangent at z_i. We use the distance r_{ij} and the angle θ_{ij} to define the shape context of the contour C. We obtain the inner distance by replacing r_{ij} with R_{ij}, the length of the shortest jointed line which joins z_i and z_j and which lies inside the closed contour.

3.3.4 Uncertainty

In choosing an appropriate common representational format we must take into account how the measurement uncertainty will propagate in a given format. The following example illustrates these concerns.

Example 3.7. Object Recognition Based on Photometric Color Invariants [5]. A simple and effective scheme for three-dimensional object recognition is to represent and match images on the basis of color histograms. For effective object recognition we should use a color space which is invariant to changes in viewing direction, object orientation and illumination.

In Table 3.1 we list several color spaces which are commonly used for this purpose (see Chapt. 16). We observe that measurement uncertainty is propagated differently in each space: the normalized rg space is unstable around $R = G = B = 0$ ($\sigma_r, \sigma_g \rightarrow \infty$) and hue H is unstable around $R = G = B$ ($\sigma_H \rightarrow \infty$) while the opponent color space o_1, o_2 is relatively stable at all RGB values [6].

Table 3.1 Photometric Invariant Color Space

Color Space	Definition	Uncertainty
Normalized rg	$r = R/S$, $g = G/S$	$\sigma_r = \sqrt{R^2(\sigma_B^2 + \sigma_G^2) + (G+B)^2\sigma_R^2/S}$.
		$\sigma_g = \sqrt{G^2(\sigma_B^2 + \sigma_R^2) + (R+B)^2\sigma_G^2/S}$.
Opponent o_1o_2	$o_1 = (R-G)/2$, $o_2 = (2B-R-G)/4$	$\sigma_1 = \sqrt{\sigma_G^2 + \sigma_R^2/2}$, $\sigma_2 = \sqrt{4\sigma_B^2 + \sigma_G^2 + \sigma_R^2/4}$.
Hue H	$\tan(H) = \sqrt{3}(G-B)/(2R-G-B)$	$\sigma_H^2 = 3(\sigma_G^2(-2BR+1) + \sigma_B^2(G^2 - 2GR)/\Delta + \sigma_R^2(1 + (G^2 - 2GB)/\Delta) + R^2\sigma_B^2\sigma_G^2)/4$.

$S = (R+G+B)$, $\Delta = R^2 + B^2 + G^2 - GR - B(G+R)^2$.

3.4 Textures

The local binary pattern (LBP) operator [2, 10] is an efficient method for representing image textures. The operator takes a local neighborhood around each pixel and thresholds the pixels in the neighborhood according to the value of the center pixel. The resulting binary valued image patch forms a normalized local texture descriptor of the image A. For a 3×3 neighborhood centered on the pixel (m,n), the LBP operator is defined as

$$LBP(m,n) = \sum_{k=0}^{7} s(A(m,n), A(i_k, j_k))2^{(k-1)} ,$$

where (i_k, j_k) are the coordinates of the kth pixel in the 3×3 neighborhood of (m,n) and

$$s(A(m,n), A(i_k, j_k)) = \begin{cases} 1 \text{ if } A(i_k, j_k) > A(m,n) , \\ 0 \text{ otherwise} . \end{cases}$$

The LBP encoding process is illustrated in Fig. 3.5.

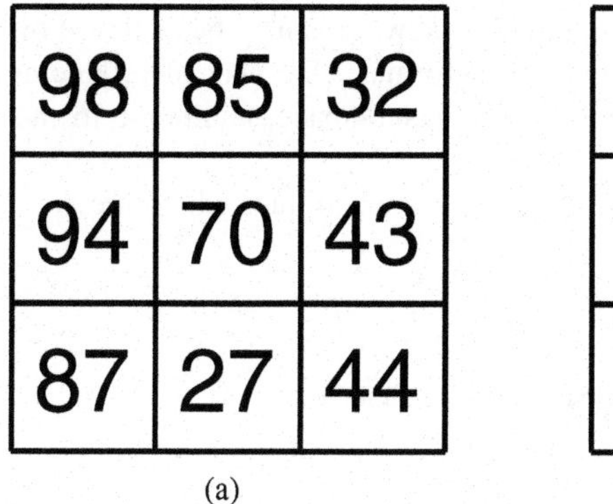

(a)

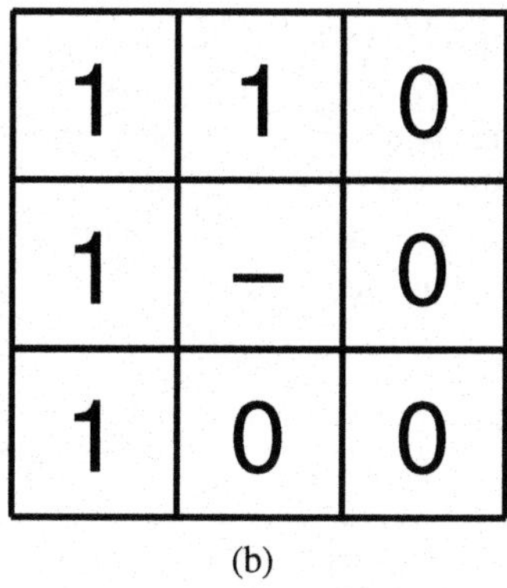

(b)

Fig. 3.5 Shows a local binary pattern operator. **(a)** Shows the gray-levels in a 3×3 neighborhood in the image A. **(b)** Shows the thresholded values. Starting from the top left-hand corner and moving in a clockwise direction, the LBP for the 3×3 local neighborhood is LBP=$\sum_{k=0}^{7} s(A(m,n), A(i_k, j_k))2^{(k-1)} = 1 \times 2^0 + 1 \times 2^1 + 0 \times 2^2 + 0 \times 2^3 + 0 \times 2^4 + 0 \times 2^5 + 1 \times 2^6 + 1 \times 2^7 = 195$.

An important extension of the LBP is the uniform LBP.

Example 3.8. Uniform Local Binary Pattern [11]. A LBP is "uniform" if it contains at most one $0-1$ and one $1-0$ transition when viewed as a circular string. For example, the LBP code in Fig. 3.5 is uniform. In a 3×3 window, only 58 of the $2^8 = 256$ patterns are uniform.

3.5 Multi-scale Representation

In some applications a multi-scale representation of the input image I is required. A wavelet decomposition (see Chapt. 8) is often used for this purpose. Another example is the SIFT key-points (Chapt. 13) which are defined as local extrema in the following multi-scale representation of I:

$$\{D(m,n|\sigma_1), D(m,n|\sigma_2), \ldots, D(m,n|\sigma_L)\} ,$$

where $D(m,n|\sigma_l)$ is the difference-of-Gaussian (DoG) representation of I at a scale σ_l:

$$D(m,n|\sigma_l) = I(m,n) \otimes G(\sigma_l) - I(m,n) \otimes G(\sigma_{l+1}) ,$$

and $I(m,n) \otimes G(\sigma_l)$ is the convolution of $I(m,n)$ with the two-dimensional zero-mean Gaussian $G(\sigma_l)$.

3.6 Sub-space Methods

In many image fusion applications an important consideration is to keep the computational load and/or the storage requirements low. This may be achieved by using a *low-dimensional* common representational format. One way of producing such a format is to apply a dimension-reducing, or sub-space, technique to the input images.

Table 3.2 lists some of the principal sub-space techniques which are commonly used for this purpose.

Table 3.2 Sub-space Techniques

Technique	Description
Principal Component Analysis (PCA)	Linear transformation chosen so the projected components have maximum variance.
Linear Discriminant Analysis (LDA)	Linear transformation for $K \geq 2$ classes. Transformation is chosen so the projected components for each class are maximally separated from the projected components of the other classes.
Independent Component Analysis (ICA)	Linear transformation chosen so the projected components have maximized independence.
Non-Negative Matrix Factorization (NMF)	Finds factors with non-negative elements.
Canonical Correlation Analysis (CCA)	For $K = 2$ finds the two transformations, one for each class, in which the projected components of the two classes are maximally correlated.

The following example illustrates the classic dimension reducing method of principal component analysis (PCA).

Example 3.9. Principal Component Analysis (PCA) [7]. The aim of principal component analysis (PCA) is to find a L-dimensional *linear* projection that best represents the input data in a least squares sense. Let the input data be K $M \times N$ input images $A_k, k \in \{1,2,\ldots,K\}$. In classical PCA we first write each A_k as a column vector $\mathbf{a}_k$:

$$\mathbf{a}_k = (a_k(1), a_k(2), \ldots, a_k(MN))^T .$$

Then we use a set of orthonormal axes $\mathbf{u}_l, l \in \{1,2,\ldots,L\}$, to obtain a L-dimensional representation of $\mathbf{a}_k$:

$$\tilde{\mathbf{a}}_k = U^T (\mathbf{a}_k - \bar{\mathbf{a}}) ,$$

where

$$U = (\mathbf{u}_1, \mathbf{u}_2, \ldots, \mathbf{u}_L) ,$$
$$\mathbf{u}_l = (u_l(1), u_l(2), \ldots, u_l(MN))^T ,$$
$$\widetilde{\mathbf{a}}_k = (\widetilde{a}_k(1), \widetilde{a}_k(2), \ldots, \widetilde{a}_k(L))^T .$$

Mathematically, the orthornormal axes $\mathbf{u}_l, l \in \{1,2,\ldots,L\}$, are given by the L dominant eigenvectors of the sample covariance matrix S:

$$S\mathbf{u}_l = \lambda_l \mathbf{u}_l ,$$

where

$$S = \frac{1}{K} \sum_{k=1}^{K} (\mathbf{a}_k - \bar{\mathbf{a}})(\mathbf{a}_k - \bar{\mathbf{a}})^T ,$$
$$\bar{\mathbf{a}} = \frac{1}{K} \sum_{k=1}^{K} \mathbf{a}_k .$$

3.7 Multiple Training Sets

A recent development in image fusion is *ensemble learning* (see Chapt. 10) in which we employ an ensemble, or collection, of *multiple* decision maps $D_k, k \in \{1,2,\ldots,K\}$, where each function D_k is learnt on its own training set T_k. Given a common (base) training set T^* we may generate an ensemble of training sets, $T_k, k \in \{1,2,\ldots,K\}$, which share the same common representational format by simply sub-sampling T^*.

Example 3.10. Bootstrapping. Given a base training set T^* of N measurements $T^*_m, m \in \{1,2,\ldots,M\}$, we create a set of K bootstrapped training sets $T_k, k \in \{1,2,\ldots,K\}$: Each bootstrapped training set T_k consists of M measurements which are selected by randomly sampling T^* with replacement. The following matlab code can be used to create a bootstrapped training set T_k.

```
for m = 1:M
    index = floor(rand(1,1) * (M - 1) + 1)
    T_k(m) = T*(index)
end
```

Table 3.3 Methods for Ensemble Sampling a Base Training Set T^*

Method	Description
Sub-sampling	Partition the base training set T^* into K disjoint slices (similar to that used in cross-validation). Each classifier S_k is trained on a training set T_k, where T_k is the base training set T^* *less* the examples in the k slice.
Bootstrapping	Perturb T^* by randomly sampling T^* with replacement. The sampling is made with a uniform probability random selection procedure. The entire procedure is repeated K times to create K different, although overlapping, training sets T_k. Each T_k contains N samples. On average each perturbed training set will have 63.2% of the samples in T^*, the rest being duplicates.
Boosting	We use the classification results obtained with the kth classifier, C_k, to learn T_{k+1}. The classifier C_k is itself learnt on T_k. The training set T_{k+1} is created by re-sampling T^* such that samples which are misclassified by S_k have a higher chance of being chosen than samples which were correctly classified by C_k.
Class Switching	Perturb T^* by randomly swapping a small percentage of class labels. The procedure is repeated K times to create K training sets $T_k, k \in \{1,2,\ldots,K\}$.

In Table 3.3 we list some methods for ensemble sampling the training set T^*.

Sometimes we require each training set T_k to have its own common representational format. This is a case of *multiple common representational formats*. Given a common (base) training set T^*, we may generate an ensemble of training sets $T_k, k \in \{1,2,\ldots,K\}$, where each T_k has a different common representational format, by applying a sub-space technique to T^* and then sub-sampling (with, or without, replacement) the result. For further details see Chapts. 10 and 11.

3.8 Software

The following matlab toolboxes are of general utility in creating a common representational format.

LIBRA. A matlab toolbox for classical and robust statistics. Authors: Sabine Verboven and Mia Hubert [16].

LBP, GETMAPPING. Two matlab m-files for the local binary pattern operator. Author: T. Ahonen [1].

MATLAB IMAGE PROCESSING, STATISTICAL AND WAVELET TOOLBOXS. Matlab toolboxes.

STPRTOOL. A statistical pattern recognition toolbox. Authors: Vojtech Franc and Vaclav Hlovac.

3.9 Further Reading

The local binary pattern (LBP) operator has been intensely investigated by the Machine Vision Group at the University of Oulu. Two important references on the subject are: [1, 11]. A modern reference which provides many pointers to the literature on the subject of invariant color spaces is [17]. For references on sub-space methods (see Chapt. 9). For object recognition we require a method for extracting two-dimensional shape enclosing contours. A modern reference on this subject is [13].

References

1. Ahonen, T.: Face and texture image analysis with quantized filter response statistics. PhD thesis, University of Oulu, Finland (2009)
2. Ahonen, T., Hadid, A., Pietikainen, M.: Face description with local binary patterns: Application to face recognition. IEEE Trans Patt. Anal. Mach. Intell. 28, 2037–2041 (2006)
3. Behloul, F., Lelieveldt, B.P.E., Boudraa, A., Janier, M., Revel, D., Reiber, J.H.C.: Neuro-fuzzy systems for computer-aided myocardial viability assessment. IEEE Trans. Medical Imag. 20, 1302–1313 (2001)
4. Belongie, S., Malik, J., Puzicha, J.: Shape matching and object recognition using shape contexts. IEEE Trans Patt. Anal. Mach. Intell. 24, 509–522 (2002)
5. Gevers, T., Smeulders, A.W.M.: Color-based object recognition. Patt. Recogn. 32, 453–464 (1999)
6. Gevers, T., Stockman, H.M.G.: Robust photometric invariant region detection in multi-spectral images. Int. J. Comp. Vis. 53, 135–151 (2003)
7. Jolliffe, I.T.: Principal Component Analysis, 2nd edn. Springer, Heidelberg (2002)
8. Ling, H., Jacobs, D.: Shape classification using the inner-distance. IEEE Trans. Patt. Anal. Mach. Intell. 29, 286–299 (2007)
9. Lowe, D.G.: Distinctive image features from scale-invariant key-points. Int. J. Comp. Vis. 60, 91–100 (2004)
10. Ojala, T., Pietikainen, M., Harwood, D.: A comparative study of texture measures with classification based on feature distributions. Patt. Recogn. 29, 51–59 (1999)
11. Ojala, T., Pietikainen, M., Maenpaa, T.: Multiresolution gray-scale and rotation invariant texture classification with local binary patterns. IEEE Trans. Patt. Anal. Mach. Intell. 24, 971–987 (2002)
12. Mikolajczek, K., Schmid, C.: A performance evaluation of local descriptors. IEEE Trans Patt. Anal. Mach. Intell. 27, 1615–1630 (2005)
13. Schlei, B.R.: A new computational framework for 2D shape-enclosing contours. Image Vis. Comp. 27, 637–647 (2009)
14. Thompson, P.M., Mega, M.S., Narr, K.L., Sowell, E.R., Blanton, R.E., Toga, A.W.: Brain image analysis and atlas construction. In: Handbook of Medical Imaging, Washington, USA, vol. 2. SPIE Press, Bellington
15. Valera Espina, M., Velastin, S.A.: Intelligent distributed surveillance systems: a review. IEE Proc. Vis. Image Sig. Process. 152, 192–204 (2005)
16. Verboven, S., Hubert, M.: Libra: A matlab library for robust analysis. Chemometrics and Intell. Laboratory Syst. 75, 127–136 (2005)
17. van de Weijer, J., Gevers, T., Smeulders, A.W.M.: Robust photometric invariant features from the color tensor. IEEE Trans. Image Process. 15, 118–127 (2006)

Chapter 4
Spatial Alignment

Abstract. The subject of this chapter is spatial alignment. In image fusion this is defined as the process of geometrically aligning two or more images of the same scene acquired at different times (multi-temporal fusion), or with different sensors (multi-modal fusion), or from different viewpoints (multi-view fusion). It is a crucial pre-processing operation in image fusion and its accuracy is a major factor in determining the quality of the output image. In order to keep our discussion focused we shall concentrate on the image registration of two input images, A and B, which we define as finding the transformation T which "optimally" maps spatial locations in the image B to the corresponding spatial locations in the image A.

4.1 Introduction

Let A and B denote two digital input images which we assume are derived from the same scene. The images will naturally have limited fields of view which will most likely be different. However, as the two images A and B are derived from the same scene we expect a relation to exist between the spatial locations in A and the spatial locations in B. If (u,v) denotes a pixel location in the reference image A and (x,y) denotes a pixel location in the floating image B, then the transformation T represents a mapping of every pixel location (x,y) in B into the corresponding location (u',v') in A [1]:

$$\begin{pmatrix} u' \\ v' \end{pmatrix} = T \begin{pmatrix} x \\ y \end{pmatrix} .$$

In general, the location (u',v') does not correspond to a pixel location in A. Let B' be the corresponding tranformed B image. The image B' is only defined at the points (u',v'), where by definition, $B'(u',v') = B(x,y)$. In order to convert $B'(u',v')$ into a digital image which is defined at the same pixel locations as A we apply an interpolation/resampling operation to $B'(u',v')$:

[1] The reader should note the subtle difference between (x,y) and (u',v'): (x,y) represents a discrete pixel location in B while (u',v') represents the corresponding floating spatial location in A. In general (u',v') does *not* correspond to a pixel location in A.

$$\widetilde{B}(u,v) \equiv R\left(B'(u',v')\right) , \tag{4.1}$$

where R is an appropriate resample/interpolation operator. In practice (4.1) is implemented by using the inverse transformation T^{-1} which maps pixels in A to their corresponding locations in B. The following example illustrates the concept of a nearest neighbor resample/interpolation operator.

Example 4.1. Nearest neighbour resample/interpolation algorithm. The simplest resample/interpolation algorithm is the nearest neighbor algorithm. Let (u,v) denote a pixel location in A. Suppose the corresponding location in B is $(x',y') = T^{-1}(u,v)$. In general (x',y') will not fall on a pixel location in B. Let $P_k = (x_k,y_k), k \in \{0,1,2,3\}$, denote the four pixel locations in B which surround the point (x',y') where P_0 is the point nearest to (x',y') (Fig. 4.1), then the nearest neighbor gray-level is $\widetilde{B}(u,v)$, where

$$\widetilde{B}(u,v) \equiv R\left(B'(u',v')\right) = B(x_0,y_0) .$$

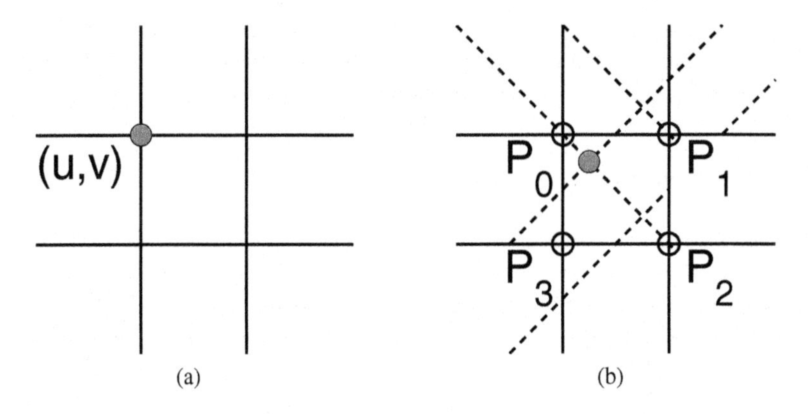

Fig. 4.1 Shows nearest neighbor interpolation. **(a)** Shows the reference image A and its grid lines as full-lines. A pixel location (u,v) is shown by a filled circle at the intersection of two grid-lines. **(b)** Shows the floating image B and its grid lines as full lines. Also shown (by dashed lines) are the inverse transformed grid lines of A. The filled circle shows the location of the inverse transformed point $(x',y') = T^{-1}(u,v)$.

4.2 Pairwise Transformation

The (pairwise) transformation

$$\begin{pmatrix} u' \\ v' \end{pmatrix} = T \begin{pmatrix} x \\ y \end{pmatrix} ,$$

Table 4.1 Spatial Transformations $T : (u',v')^T = T\left((x,y)^T\right)$

Name	Formula
Translation	$u' = x + a_1, v' = y + a_2.$
Similarity	$u' = a_1x + a_2y + a_3,\ v' = -a_2x + a_1y + a_4.$
Affine	$u' = a_1x + a_2y + a_3,\ v' = a_4x + a_5y + a_6.$
Perspective	$u' = (a_1x + a_2y + a_3)/(a_7x + a_8y + 1),$ $v' = (a_4x + a_5y + a_6)/(a_7x + a_8y + 1).$
Polynomial	$u' = \sum a_{ij}x^iy^j,\ v' = \sum b_{ij}x^iy^j.$

is a mathematical relationship that maps a spatial location $(x,y)^T$ in one image to a new location, $(u',v')^T$, in another image. The choice of transformation is always a compromise between a smooth distortion and a distortion which achieves a good match. One way to ensure smoothness is to assume a low-order parametric form for the transformation [8, 23] such as that given in Table 4.1. In most applications the transformation T is chosen on the grounds of mathematical convenience. However, sometimes, we may have information regarding the physical processes which govern the formation of the pictures. In this case we may use physical arguments in order to *derive* the transformation T.

In many applications, the images also undergo local deformations. In this case, we cannot describe the alignment of two images using a single low-order transformation. In this case we use a composite transformation T, which consists of a low-order global transformation T_G and a local transformation T_L:

$$\begin{pmatrix} u' \\ v' \end{pmatrix} = T\begin{pmatrix} x \\ y \end{pmatrix} = T_G\begin{pmatrix} x \\ y \end{pmatrix} + T_L\begin{pmatrix} x \\ y \end{pmatrix},$$

where the parameters of T_L change with (x,y).

The thin-plate spline (TPS) is often used to model the composite transformation T.

4.2.1 Thin-Plate Splines

Mathematically, the TPS model for the composite transformation T is:

$$u' = a_1 + a_2x + a_3y + \sum_{m=1}^{M} \alpha_m r_m^2 \ln r_m^2 ,$$

$$v' = a_4 + a_5x + a_6y + \sum_{m=1}^{M} \beta_m r_m^2 \ln r_m^2 ,$$

where $(x_m,y_m), m \in \{1,2,\ldots,M\}$, is a set of known anchor points and $r_m^2 = (x - x_m)^2 + (y - y_m)^2 + d^2$. Apart from the parameter d, the transformation T has six parameters, $a_1,a_2,\ldots,a_6$, corresponding to the global affine transformation T_G and

$2N$ parameters $(\alpha_m, \beta_m), m \in \{1,2,\ldots,M\}$, corresponding to the local transformation T_L, and which satisfy the following constraints:

$$\sum_{m=1}^{M} \alpha_m = 0 = \sum_{m=1}^{M} \beta_m \,,$$
$$\sum_{m=1}^{M} x_m \alpha_m = 0 = \sum_{m=1}^{M} x_m \beta_m \,,$$
$$\sum_{m=1}^{N} y_m \alpha_m = 0 = \sum_{m=1}^{M} y_m \beta_m \,.$$

The TPS coefficients can be calculated using a least square solution [12]:

$$\begin{pmatrix} \alpha_1 & \beta_1 \\ \alpha_2 & \beta_2 \\ \vdots & \vdots \\ \alpha_M & \beta_M \\ a_2 & b_2 \\ a_3 & b_3 \\ a_1 & b_1 \end{pmatrix} = \begin{pmatrix} 0 & U(r_{12}) & \ldots & U(r_{1M}) & 1 & x_1 & y_1 \\ U(r_{21}) & 0 & \ldots & U(r_{2M}) & 1 & x_2 & y_2 \\ \vdots & \vdots & \ddots & \vdots & \vdots & \vdots & \vdots \\ U(r_{M1}) & U(r_{M2}) & \ldots & 0 & 1 & x_M & y_M \\ 1 & 1 & \ldots & 1 & 0 & 0 & 0 \\ x_1 & x_2 & \ldots & x_M & 0 & 0 & 0 \\ y_1 & y_2 & \ldots & y_M & 0 & 0 & 0 \end{pmatrix}^{-1} \begin{pmatrix} u'_1 & v'_1 \\ u'_2 & v'_2 \\ \vdots & \vdots \\ u'_M & v'_M \\ 0 & 0 \\ 0 & 0 \\ 0 & 0 \end{pmatrix},$$

where $r_{ij}^2 = (x_i - u'_j)^2 + (y_j - v'_j)^2 + d^2$ and $U(r) = r^2 \ln r^2$. For further details concerning the estimation of the TPS parameters see [22].The following example illustrates the use of a TPS to model the warping of a fingerprint.

Example 4.2. Fingerprint Warping Using a Thin-plate spline [18]. The performance of a fingerprint matching system is affected by the nonlinear deformations introduced in the fingerprint during image acquisition This nonlinear deformation is represented using a global affine transformation T_G and a local transformation T_L. We use a TPS function [22] to represent the composite transformation:

$$u' = a_1 + a_2 x + a_3 y + \sum_{m=1}^{M} \alpha_m r_m^2 \ln r_m^2 \,,$$
$$v' = a_4 + a_5 x + a_6 y + \sum_{m=1}^{M} \beta_m r_m^2 \ln r_m^2 \,,$$

where $(x_m, y_m), m \in \{1,2,\ldots,M\}$, is a set of known anchor points and $r_m^2 = (x - x_m)^2 + (y - y_m)^2 + d^2$.

4.3 Hierarchical Registration

The simplest approach to register two images A and B and to calculate the transformation T is to decompose T into numerous local affine registrations of small sub-images. The idea is to reduce the complexity of the registration process using a hierarchical strategy (see Fig. 4.2).

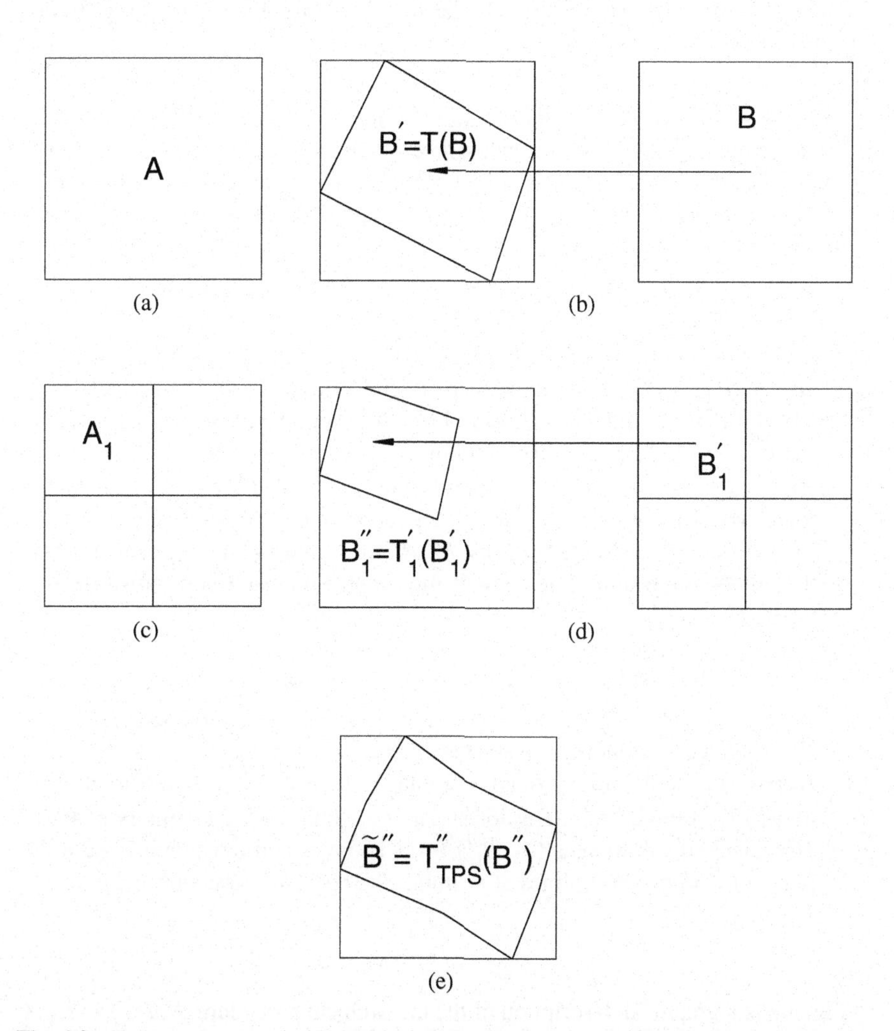

Fig. 4.2 Shows two stages in hierarchical registration process of the floating image B with the reference image A. **(a)** Shows the reference image A. **(b)** Shows the input floating image B and the transformed floating image $B' = T(B)$ after the first stage in global registration. **(c)** Shows the reference image divided into 4 quadrants. **(d)** Shows the transformed image B' divided into four quadrants and the transformed quadrants $B''_k = T'(B'_k)$. (e) Shows the composite image $\widetilde{B}'' = T''_{TPS}(B'')$ formed by applying the TPS transformation to the transformed quadrants $B''_k, k \in \{1,2,\ldots,4\}$.

In the hierarchical approach [1, 13] we progressively subdivide the input images A and B into smaller sub-images which are separately registered by maximizing an appropriate image similarity measure. We separately model each sub-image registrations with an affine transformation. Then the final composite transformation T is found by assimilating all the sub-image transformations using a TPS interpolation algorithm.

The following example explains the basic steps in the hierarchical registration scheme of Likar-Pernus [13].

Example 4.3. Likar-Pernus Hierarchical Registration Scheme [13]. In the Likar-Pernus algorithm we progressively subdivide the two input images A and B. We automatically register the sub-images, and then apply the thin-plate splines interpolation between the centers of registered sub-images. The steps for the first two hierarchical levels are:

1. Register B to the reference image A using an affine transformation T. Let $B' = T(B)$ be the transformed image.
2. Separately partition the images A and B' into four sub-images A_1, A_2, A_3, A_4 and B'_1, B'_2, B'_3, B'_4 of identical size. Each corresponding sub-image pair (A_k, B'_k) is independently registered by using an affine transformation T'_k. Let $B''_k = T'_k(B'_k)$ be the corresponding transformed sub-image.
3. Assimilate the four transformed sub-images $B''_1, B''_2, B''_3, B''_4$ into a single transformed image $\widetilde{B}''$ as follows: The coordinates of the centers of the four registered sub-images $B''_1, B''_2, B''_3, B''_4$ form four point pairs, which are the inputs to the thin-plate splines algorithm. The result is a transformed image $\widetilde{B}'' = T''_{TPS}(B'')$.
4. Separately partition the registered images A and $\widetilde{B}''$ into 16 sub-images $A_1, A_2, \ldots, A_{16}$ and $\widetilde{B}''_1, \widetilde{B}''_2, \ldots, \widetilde{B}''_{16}$ of identical size. Each corresponding sub-image pair $(A_k, \widetilde{B}''_k)$ is registered using an affine transformation T''_k. Let B''' be the corresponding transformed image.
5. Assimilate the 16 transformed sub-images $B'''_1, B'''_2, \ldots, B'''_{16}$ into a single transformed image $\widetilde{B}'''$ as follows: The coordinates of the centers of the 16 registered sub-images (A_k, B'''_k) form 16 point pairs, which are the inputs to the thin-plate splines algorithm. The result is a transformed image $\widetilde{B}''' = T'''_{TPS}(B''')$.

Fig. 4.2 is a graphical description of the hierarchical procedure. *Note*: The registration of finer details is preceded by registration and smooth interpolation obtained at a more global scale.

In principle, the hierarchical decomposition of the images A and B may be continued until the sub-images contain only one pixel. However, in practice, we stop the decomposition process much earlier. The reasons for this are twofold: (1) The algorithm is sensitive to the accuracy of the sub-image registrations: a misregistration at a given hierarchical level can propagate to the lower hierarchical levels. (2)

The probability of a misregistration increases as we move to the lower hierarchical levels. This is because, in general, the reliability of an image similarity measure decreases with the size of the image patch (Chapt 14).

One way to prevent registration errors propagating down the hierarchy is to *adaptively* stop the hierarchical sub-division before the image patches become so small they are effectively "structureless". For this purpose we may use the following test: An image patch (containing K pixels, $(x_k, y_k), k \in \{1, 2, \ldots, K\}$, with gray-levels g_k) is said to be structureless if $\rho > \tau$, where ρ is Moran's autocorrelation coefficient [1]:

$$\rho = \frac{K \sum_{h,k} w_{hk}(g_h - \bar{g})(g_k - \bar{g})}{\sigma \sum_{h,k} w_{hk} \sum_h (g_h - \bar{g})^2} ,$$

τ is a given threshold and $w_{hk} = 1/\sqrt{(x_k - x_h)^2 + (y_k - y_h)^2}$ is the inverse Euclidean distance between (x_k, y_k) and (x_h, y_h).

4.4 Mosaic Image

Thus far we have considered the problem of registering a pair of images. In some applications we are interested in building a single panoramic or "mosaic" image $\widetilde{I}$ from multiple images $I_k, k \in \{1, 2, \ldots, K\}$. To do this we need to find functions T_k which transform each input image I_k onto the image $\widetilde{I}$.

Building a mosaic image from a sequence of partial views is a powerful means of obtaining a broader view of a scene than is available with a single view. Research on automated mosaic construction is ongoing with a wide range of different applications.

Example 4.4. Mosaic Fingerprint Image [9]. Fingerprint-based verification systems have gained immense popularity due to the high level of uniqueness attributed to fingerprints and the availability of compact solid-state fingerprint sensors. However, the solid-state sensors sense only a limited portion of the fingerprint pattern and this may limit the accuracy of the user verification. To deal with this problem we may construct a mosaic fingerprint image from multiple fingerprint impressions.

Example 4.5. Mosaic Image of the Retina [3, 4]. One area in which mosaic images are particularly valuable is in the diagnosis and treatment of diseases of the retina. A seamless mosaic image which is formed from multiple fundus camera images aids in the diagnosis and provides a means for monitoring the progression of different diseases. It may also be used as a spatial map of the retina during surgical treatment.

At first sight we may assume that the ability to spatially align a pair of images is sufficient to solve the problem of forming a mosaic of the entire scene from multiple partial views. Theoretically, if one image can be established as an "anchor image" I_0 on which to base the mosaic image $\tilde{I}$, then the transformation of each remaining image onto this anchor may be estimated using pairwise registration. The mosaic image I^* is then formed by "stitching" together the transformed images $T_m(I_m)$. Unfortunately, in practice, this approach may not work for the following reasons:

Non-Overlap. Some images may not overlap with the anchor image at all. This makes a direct computation of the transformation impossible. In other cases, images may have insufficient overlap with the anchor image to compute a stable transformation. The straightforward solution is to compose transformations using an "intermediate" image. This is problematic, however, since repeated application of the transformation will often magnify the registration error.

Inconsistent T_m. The transformations T_m may be mutually inconsistent. This may happen even if all the image-to-anchor transformations have been accurately estimated. The reason for this is as follows: Although each image may individually register accurately with the anchor image and the non-anchor images may register accurately with each other, this does not ensure that the transformations onto the anchor image are *mutually consistent*.

One approach to solving this problem is to constrain the transformations so that they are all mutually consistent.

Example 4.6. Transformation Constraints in a Mosaic Image. Given a sequence of N images $I_1, I_2, \ldots, I_N$, we estimate $N(N-1)$ pairwise transformations

$$\begin{pmatrix} u' \\ v' \end{pmatrix} = T_{ij} \begin{pmatrix} x \\ y \end{pmatrix} ,$$

where $(u', v')^T$ and $(x, y)^T$ denote, respectively, the coordinates of corresponding points in I_i and I_j. The T_{ij} must satisfy the following relationships:

$$T_{ik} = T_{ij} \circ T_{jk} ,$$
$$T_{ij} = T_{ji}^{-1} ,$$

where $T_{ij} \circ T_{jK}$ denotes the application of T_{jk} followed by the application of T_{ij}.

For an affine transformation, the transformation T_{ij} can be written in matrix form as $\begin{pmatrix} u' \\ v' \end{pmatrix} = A_{ij} \begin{pmatrix} x \\ y \end{pmatrix} + B_{ij}$. In this case, the above relationships become

$$A_{ik} = A_{ij} A_{jk} ,$$
$$B_{ik} = A_{ij} B_{jk} + B_{ij} .$$

4.4.1 Stitching

In "stitching" together the transformed images $T_m(I_m)$ our aim is to produce a visually plausible mosaic image $\tilde{I}$ in which, I^* is geometrically and photometrically as similar as possible to the input images $T_m(I_m)$ and the seams between the stitched images are invisible. The stitching algorithms which are commonly used nowadays fall into two types:

Optimal Seam Algorithms. These algorithms search for a curve in the overlap region on which the differences between the $T_m(I_m)$ are minimized. Then each image is copied to the corresponding side of the seam.

Transition Smoothing Algorithms. These algorithms minimize seam artifacts by smoothing the transition region (a small region which is spatially near to the seam) [24].

The following example describes the transition smoothing, or feathering, of two input images I_1 and I_2.

Example 4.7. Feathering [24]. In feathering, the mosaic image $\tilde{I}$ is a weighted combination of the input images I_1, I_2, where the weighting coefficients vary as a function of the distance from the seam. In general, feathering works well as long as there is no significant misalignment. However, when the misalignments are significant, the mosaic image displays artifacts such as double edges. A modification of feathering which is less sensitive to misalignment errors, is to stitch the derivatives of the input images instead of the images themselves. Let $\partial I_1/\partial x$, $\partial I_1/\partial y$, $\partial I_2/\partial x$ and $\partial I_2/\partial y$ be the derivatives of the input images. If F_x and F_y denote the derivative images formed by feathering $\partial I_1/\partial x$ and $\partial I_2/\partial x$ and $\partial I_1/\partial y$ and $\partial I_2/\partial y$, then we choose the final mosaic image $\tilde{I}$ to be the image whose derivatives $\partial \tilde{I}/\partial x$ and $\partial \tilde{I}/\partial y$ are closest to F_x and F_y.

4.5 Image Similarity Measures

In order to be able to register two images, a measure has to be defined to numerically quantify the goodness of fit between the images, namely the similarity measure. The choice of the appropriate similarity measure is crucial for a successful image registration procedure, so the decisive criterion is the type of images to be registered. Therefore depending on the type of the modalities used to acquire the images, the user can choose between several similarity measures (see Chapt. 14). In this chapter we shall concentrate on the mutual information similarity measure. This has been found to be the most successful especially when the input images are heterogeneous, i. e. they were captured with different sensors or with different spectral bands or with different spatial resolutions [7].

4.6 Mutual Information

Given a reference image A and a spatially aligned and resampled image B [2], the mutual information of A and B is defined as

$$MI(A,B) = \int\int p_{AB}(a,b)\log_2 \frac{p_{AB}(a,b)}{p_A(a)p_B(b)} dxdy\,, \tag{4.2}$$

where $p_A(a)$ is the probability a pixel (x,y) in A has a gray-level a, $p_B(b)$ is the probability a pixel (x,y) in B has a gray-level b and $p_{AB}(a,b)$ is the joint probability a pixel (x,y) in A has a gray-level a and the same pixel in B has a gray-level b.

4.6.1 Normalized Mutual Information

The integral in (4.2) is taken over the pixels which are common to both A and B. As a result, $MI(A,B)$ may vary if the number of pixels which are common to A and B changes. In general the variations in $MI(A,B)$ are small but they may lead to inaccuracies in a spatial alignment algorithm. To avoid these inaccuracies, we often use a normalized mutual information similarity measure in place of $MI(A,B)$. Four commonly used normalized MI measures are [10]:

$$NMI(A,B) = \begin{cases} \dfrac{MI(A,B)}{H(A)+H(B)} \\ \dfrac{MI(A,B)}{\min\left(H(A),H(B)\right)}\,, \\ \dfrac{MI(A,B)}{H(A,B)}\,, \\ \dfrac{MI(A,B)}{\sqrt{H(A)H(B)}}\,, \end{cases}$$

where

$$\begin{aligned} H(A) &= -\int\int p_A(a)\log_2 p_A(a)dxdy\,, \\ H(B) &= -\int\int p_B(b)\log_2 p_B(b)dxdy\,, \\ H(A,B) &= -\int\int p_{AB}(a,b)\log_2 p_{AB}(a,b)dxdy\,, \end{aligned}$$

and the integration is performed over the overlap region of the images A and B.

---

[2] *Note*: In this section B denotes the spatially aligned and re-sampled floating image.

4.6.2 Calculation

The most common approach to calculate the mutual information $MI(A,B)$ and the normalized mutual information $NMI(A,B)$ is to calculate the entropies $H(A)$, $H(B)$ and $H(A,B)$ using the marginal probabilities $p_A(a)$ and $p_B(b)$ and the joint probability $p_{AB}(a,b)$. Since, the marginal probabilities may be derived from $p_{AB}(a,b)$:

$$p_A(a) = \int p_{AB}(a,b)db \,, \qquad p_B(b) = \int p_{AB}(a,b)da \,,$$

we need only consider the calculation of $p_{AB}(a,b)$.

4.6.3 Histogram

The most straightforward way to calculate the joint probability distribution $p_{AB}(a,b)$ is to use a discrete histogram H_{AB} as follows: We quantize the gray-levels in A and B into P and Q bins respectively Then we approximate $p_{AB}(a,b)$ using the two-dimensional histogram $H_{AB} = (h_{AB}(1,1), h_{AB}(1,2), \ldots, h_{AB}(P,Q))^T$, where $h_{AB}(p,q)$ is the number of pixels whose gray-levels in A fall in the pth bin and whose gray-levels in B fall in the qth bin.

In this case, the formula for the mutual information, is

$$MI(A,B) = \sum_{(p,q)} h_{AB}(p,q) \log_2 \left(\frac{h_{AB}(p,q)}{h_A(p) h_B(q)} \right) \Big/ \sum_{(p,q)} h_{AB}(p,q) \,,$$

where

$$h_A(p) = \sum_q h_{AB}(p,q) \qquad \text{and} \qquad h_B(q) = \sum_p h_{AB}(p,q) \,.$$

Although widely used, the histogram method suffers from several drawbacks: It yields a discontinuous density estimate and there is no principled method for choosing the size and placement of the bins. For example, if the bin width is too small, the density estimate is noisy while if the bin width is too large the density estimate is too smooth. Legg *et al.* [14] recommends using Sturges' rule for the optimal bin width:

$$w = \frac{r}{1 + \log_2(K)} \,,$$

where r is the range of gray-level values, K is the number of elements in the input image. In this case the optimal number of bins is r/w.

A partial solution to these problems is to calculate $p_{AB}(a,b)$ using the method of Parzen windows.

4.6.4 Parzen Windows

Instead of using discrete histogram bins to calculate the joint probability distribution $p_{AB}(a,b)$, we use continuous bins. This is known as kernel, or Parzen-window, density estimation [19, 20] and is a generalization of histogram binning. If A and B each contain K pixels with gray-levels $a_k, b_k, k \in \{1,2,\ldots,K\}$, then the estimated joint density $p_{AB}(a,b)$ is given by

$$p_{AB}(a,b) = \frac{1}{K^2\sigma_A\sigma_B}\sum_{k=1}^{K}\sum_{l=1}^{K} H\left(\frac{a-a_k}{\sigma_A}\right)H\left(\frac{b-b_l}{\sigma_B}\right),$$

where $H(x)$ denotes a kernel function which satisfies $\int_x H(x)dx = 1$. In general a density estimate $p(x)$ is more sensitive to the choice of the bandwidth σ and less sensitive to the choice of the kernel $H(x)$. For this reason we often use a zero-mean Gaussian function with standard deviation σ for the kernel $H(x)$. Table 4.2 lists several schemes which are commonly used to calculate the optimal bandwidth σ.

Table 4.2 Methods for Calculating Optimum One-Dimensional Bandwidth σ

Name	Description
Rule-of-Thumb	Suppose the input data (consisting of N measurements $a_i, i \in \{1,2,\ldots,N\}$), is generated by a given parametric density function, e. g. a Gaussian function. In this case $\sigma = 1.06\hat{\sigma}N^{-1/5}$, where $\hat{\sigma}$ is the sample standard deviation. Robust versions of this bandwidth are available: $\sigma = 1.06\min(\hat{\sigma}, \hat{Q}/1.34)N^{-1/5}$ and $\sigma = 1.06\hat{s}N^{-1/5}$, where $\hat{Q}$ is the sample interquartile distance and $\hat{s} = \mathrm{med}_j\vert a_j - \mathrm{med}_i a_i\vert$.
Cross-Validation (CV)	Use a CV procedure to directly minimize the MISE or the AMISE. CV variants include least square, biased and smoothed CV [11].
Plug-in	Minimize the AMISE using a second bandwidth known as the *pilot* bandwidth Σ. In the *solve-the-equation plug-in* method we write L as a function of the kernel bandwidth σ [11].

MISE is the mean integrated square error and is defined as $\mathrm{MISE}(p,\hat{p}_\sigma) = \int (p(a) - \hat{p}_\sigma(a))^2 da$, where $\hat{p}_\sigma(a)$ is the kernel approximation to $p(a)$. AMISE denotes the asymptotic MISE and represents a large number approximation of the MISE.

4.6.5 Iso-intensity Lines

Iso-intensity lines [17] is a new scheme developed specifically for calculating the joint probability density $p_{AB}(a,b)$. Suppose the gray-levels in A and B are quantized, respectively, into P and Q bins. For each pixel location (m,n) we estimate the gray-level values G_1, G_2, G_3, G_4 of its four neighbors which lie at a horizontal or vertical distance of half a pixel from (m,n). We divide the square defined by these neighbors into a pair of triangles (see Fig. 4.3). Within the triangle we suppose the gray-level values vary linearly as follows:

$$A(m+\delta x, n+\delta y) = a_A x + b_A y + c_A \ ,$$
$$B(m+\delta x, n+\delta y) = a_B x + b_B y + c_B \ ,$$

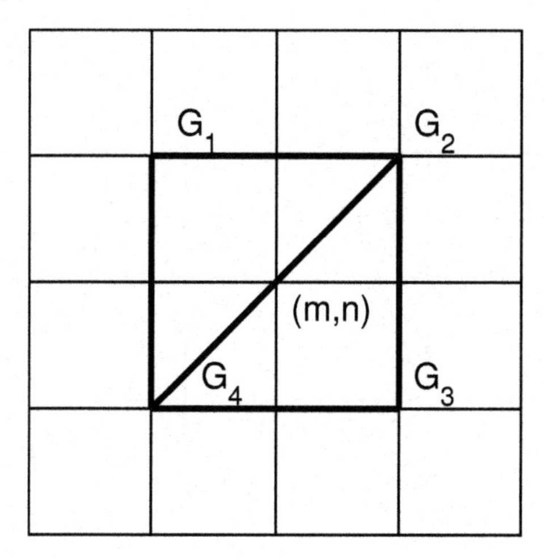

Fig. 4.3 Shows the pixel (m,n) with the gray-levels $G_k, k \in \{1,2,3,4\}$

where $A(m+\delta x, n+\delta y)$ and $B(m+\delta x, n+\delta y)$ denote, respectively, the gray-level of a point $(m+\delta x, n+\delta y)$ in the triangle and $-0.5 \leq \delta x, \delta y \leq 0.5$. To calculate the joint distribution of the two images A and B we sequentially consider the PQ different gray-level pairs denoted as (α, β). For each pixel (m,n) we see whether the pair of corresponding triangles contains a point $(m+\delta x, n+\delta y)$ which has a gray-level value α in A and β in B. Such a point $(m+\delta x, n+\delta y)$ contributes a vote to the entry (α, β) in $p_{AB}(a,b)$.

4.7 Partial Volume Interpolation

The histogram, Parzen and iso-intensity line algorithms all assume the images A and B are spatially aligned and if necessary image interpolation has been performed. The partial volume interpolation (PVI) is an alternative technique which does not assume spatial alignment or image interpolation [16]. It works as follows.

Suppose T represents a mapping of the pixel (x,y) in B into the corresponding location (u',v') in A. In general (u',v') will not correspond to a pixel location in A. Suppose $Q_k = (u_k, v_k), k \in \{0,1,2,3\}$, are the four pixel locations in A which surround (u',v'). (Fig. 4.4). Then if $A(u_k, v_k)$ has a quantized gray-level α_k and $B(x,y)$ has a quantized gray-level β, then $H_{AB}(\alpha_k, \beta)$ receives a fractional vote equal to

$$r_k^{-1} / \sum_{h=0}^{3} r_h^{-1} ,$$

where $r_k = \sqrt{(u_k - u')^2 + (v_k - v')^2}$.

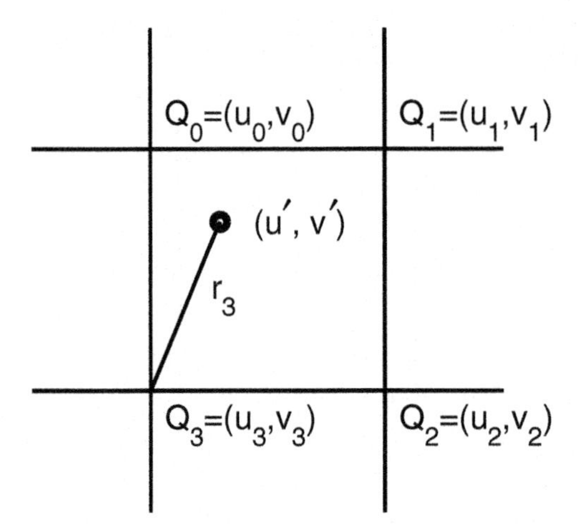

Fig. 4.4 Shows the four pixel locations $Q_k = (u_k, v_k), k \in \{0, 1, \ldots, 3\}$, in A which surround the transformed point (u', v'). Also shown is the Euclidean distance $r_3 = \sqrt{(u_3 - u')^2 + (v_3 - v')^2}$ from Q_3 to (u', v').

4.8 Artifacts

The success of the mutual information algorithms image registration lie in their inherent simplicity. It makes few assumptions regarding the relationship that exists between different images. It only assumes a *statistical* dependence. The idea is that although different sensors may produce very different images, since they are imaging the same underlying scene, there will exist some inherent mutual information between the images. When the images (or image patches) are spatially aligned, then the mutual information is maximal. To be an effective similarity measure, however, we require the mutual information to fall monotonically to zero as we move away from perfect alignment. In practice, the MI does not fall monotonically to zero.

These artifacts are due to inaccuracies in estimating the marginal densities $p_A(a)$ and $p_B(b)$ and the joint density $p_{AB}(a, b)$. The artifacts are of two types:

Interpolation effects [21]. Initially, when the images are aligned the pixel locations of A and B coincide. Therefore no interpolation is needed when estimating the joint intensity histogram. At the same time the dispersion of the histogram is minimal when the images are registered and therefore the joint entropy is minimal. By translating the floating image B with an integer number of the pixel dimension, the grid points of the two images will again be aligned avoiding the need for interpolation, but the dispersion of the joint histogram is increasing due to misregistration, reducing the MI accordingly. For all other translations, corresponding to some fraction of pixel dimension, the pixel locations of the images do not coincide anymore and therefore interpolation is required to estimate intensity values between pixel locations of the reference image. As a consequence the joint histogram is not only dispersed because of the image content and a possible misregistration, but it also contains an additional dispersion induced by the interpolation method. More dispersion implies a higher joint entropy value, which in turn decreases the MI of the reference image between the pixel location. The MI is found to vary as shown in Fig.4.5. An effective way to reduce the interpolation effects is to use nearest neighbor interpolation with jittered sampling [21]. We jitter the coordinates of each pixel which is to be interpolated by adding a normally distributed random offset (zero mean and standard deviation of one-half).

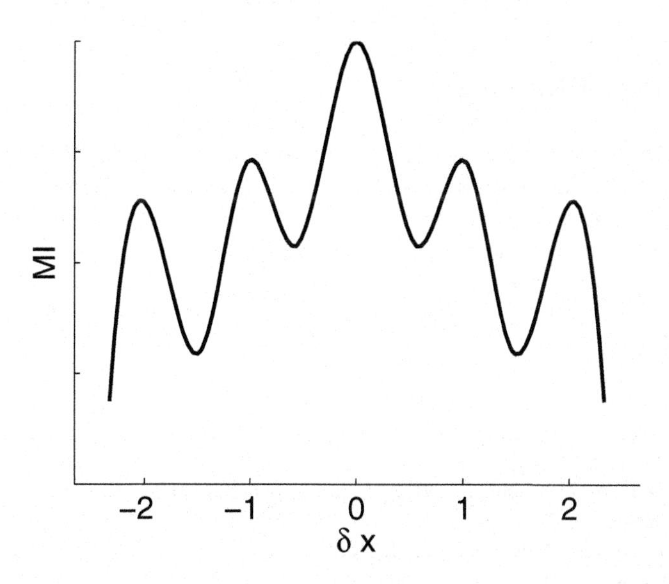

Fig. 4.5 Shows the typical interpolation artifacts as a function of the relative displacement δx between the reference image and the floating image. *Note*: PVI interpolation curves are often concave in shape.

Small size effects. Sometimes an image patches with a low structural content may appear. This often occurs when we register the two images using a hierarchical matching algorithm. These structureless patches often lead to inconsistent local registrations due to a low MI response. If two signals are independent then their MI reaches its minimum possible value of zero. We might expect, therefore, that by shifting a structureless sub-image around its initial position, the similarity measure will have a small response. Surprisingly this is not true. The MI starts to increase as soon as a structureless sub-image overlaps a region of higher structural content. One explanation for this phenomena is the following [1]: The number of samples required to obtain a consistent estimate of the marginal entropies $H(A)$ and $H(B)$ is much less than the number of samples required to obtain a consistent of the joint entropy $H(A,B)$.

4.9 Software

KDE, KDE2D. Automatic data-driven bandwidth selection functions. Available from the matlab central depository. Author: Zdravko Botev [2].

MATLAB IMAGE PROCESSING TOOLBOX. Matlab image processing toolbox. The toolbox contains m-files for performing image registration, re-sampling and interpolation.

THIN PLATE SPLINES. Suite of matlab m-files for performing thin plate spline interpolation. Available from matlab central depository. Author: Bing Jian.

4.10 Further Reading

The subject of image registration has been intensely investigated for many years. A modern review of image warping is [8]. Specific references on the use of mutual information for image registration are [15, 16]. The calculation of mutual information has been considered by many authors including [5, 6].

References

1. Andronache, A., von Siebenthal, M., Szekely, G., Cattin, P.: Non-rigid registration of multi-modal images using both mutual information and cross-correlation. Med. Imag. Anal. 12, 3–15 (2006)
2. Botev, Z.I.: A novel nonparametric density estimator. Technical Report. The University of Queensland
3. Can, A., Stewart, C.V., Roysam, B., Tanenbaum, H.L.: A feature-based technique for joint, linear estimation of higher-order image-to-mosaic transformations: mosaicing the curved human retina. IEEE Trans Patt. Anal. Mach. Intell. 24, 412–419 (2002)
4. Can, A., Stewart, C.V., Roysam, B., Tanenbaum, H.L.: A feature-based, robust, hierarchical algorithm for registering pairs of images of the curved human retina. IEEE Trans. Patt. Anal. Mach. Intell. 24, 347–364 (2002)
5. Darbellay, G.A.: An estimator for the mutual information based on a criterion for independence. Comp. Stats. Data Anal. 32, 1–17 (1999)
6. Darbellay, G.A., Vajda, I.: Estimation of the information by an adaptive partitioning of the observation space. IEEE Trans. Inf. Theory 45, 1315–1321 (1999)
7. Fransens, R., Strecha, C., van Gool, L.: Multimodal and multiband image registration using mutual information. In: Proc. ESA-EUSC (2004)
8. Glasbey, C.A., Mardia, K.V.: A review of image warping methods. J. Appl. Stat. 25, 155–171 (1998)
9. Jain, A., Ross, A.: Fingerprint mosaicking. In: IEEE Int. Conf. ICASSP (2002)
10. Hossny, M., Nahavandi, S., Creighton, D.: Comments on Information measure for performance of image fusion. Elect. Lett. 44, 1066–1067 (2008)
11. Jones, M.C., Marron, J.S., Sheather, S.J.: A brief survey of bandwidth selection for density estimation. J. Am. Stat. Assoc. 91, 401–407 (1996)
12. Likar, B., Pernus, F.: Registration of serial transverse sections of muscle fibers. Cytometry 37, 93–106 (1999)

13. Likar, B., Pernus, F.: A hierarchical approach to elastic registration based on mutual information. Image Vis. Comp. 19, 33–44 (2001)
14. Legg, P.A., Rosin, P.L., Marshall, D., Morgan, J.E.: Improving accuracy and efficiency of registration by mutual information using Sturges' histogram rule. In: Proc. Med. Image Understand. Anal., pp. 26–30 (2007)
15. Maes, F., Vandermeulen, D., Suetens, P.: Medical image registration using mutual information. Proc. IEEE 91, 1699–1722 (2003)
16. Pluim, J.P.W., Maintz, J.B.A., Viergever, M.A.: Mutual information based registration of medical images: a survey. IEEE Trans. Med. Imag. 22, 986–1004 (2003)
17. Rajwade, A., Banerjee, A., Rangarajan, A.: Probability density estimation using isocontours and isosurfaces: application to information theoretic image registration. IEEE Trans. Patt. Anal. Mach. Intell. (2009)
18. Ross, A., Dass, S.C., Jain, A.K.: A deformable model for fingerprint matching. Patt. Recogn. 38, 95–103 (2005)
19. Scott, D.W.: Multivariate Density Estimation. Wiley, Chichester (1992)
20. Silverman, B.: Density Estimation for Statistical Data Analysis. Chapman and Hall, Boca Raton (1986)
21. Tsao, J.: Interpolation artifacts in multimodality image registration based on maximization of mutual information. IEEE Trans. Med. Imag. 22, 854–864 (2003)
22. Zagorchev, L., Goshtasby, A.: A comparative study of transformation functions for nonrigid image registration. IEEE Trans. Image Process. 15, 529–538 (2006)
23. Zitova, B., Flusser, J.: Image registration: A survey. Image Vis. Comput. 21, 977–1000 (2003)
24. Zomet, A., Levin, A., Peleg, S., Weiss, Y.: Seamless image stitching by minimizing false edges. IEEE Trans. Image Process. 15, 969–977 (2006)

Chapter 5
Semantic Equivalence

Abstract. The subject of this chapter is semantic equalization. This is the conversion of input data which does not refer to the same object or phenomena to a common object or phenomena. Different inputs can only be fused together if they refer to the same object or phenomena. In the case of image fusion we normally assume this to be the case if the images are captured by the same or similar type of camera. However, in the case of feature map fusion, the feature maps rarely refer to the same object or phenomena. In this case, fusion can only take place if the features maps are semantically equivalent. This is also true in the case of decision map fusion. In this chapter we shall therefore concentrate on the semantic equivalence of feature maps and decision maps.

5.1 Introduction

In order to carry out image fusion, feature map fusion or decision label fusion, we must first ensure the input data which is to be fused together is semantically equivalent. Two items are are said to be semantically equivalent if (1) they refer to the same object or phenomena or (2) they do not refer to the same object or phenomena but are causally linked to a common object or phenomena.

Although in principle the issue of semantic equivalence affects image fusion, feature map fusion and decision label fusion, in practice we generally assume the input images in image fusion applications are semantically equivalent. This is especially true if the input images are captured by the same type of camera. However, feature maps and to a less extent, decision labels are rarely semantically equivalent and fusion cannot take place without conversion to a common object or phenomena. This process is known as semantic equalization and forms the subject of this chapter.

The following two examples illustrate the concept of semantic equivalence of two feature maps A and B.

Example 5.1. Multiple Edge Maps. Consider an input image on which we separately apply a Sobel edge detector and a Canny edge detector. The two detectors work on different principles but both measure the presence, or absence, of an edge in the input image. The corresponding two feature maps, F_{sobel} and F_{canny}, are therefore semantically equivalent. In fact, if F_{sobel} and F_{canny} use the same radiometric scale, then we may fuse them together without any further processing.

Example 5.2. Target Detection. Consider an input image on which we wish to test for the presence, or otherwise, of a target. We apply an edge operator and a blob operator to the input image. The two detectors work on different principles and both measure very different characteristics of the image. In this case, the corresponding feature maps, F_{edge} and F_{blob}, do not refer to the same object or phenomena and are not, therefore, semantically equivalent.

However, according to the theory of target detection, both F_{edge} and F_{blob} are causally linked to the presence, or absence, of a target. In this case, we may semantically align the two feature by converting $F_{edge}(x,y)$ and $F_{blob}(x,y)$ into evidence that a target is present at (x,y). If we use the same evidence scale for both F_{edge} and F_{blob}, then radiometric calibration is not required and we may fuse F_{edge} and F_{blob} together without any further processing.

5.2 Probabilistic Scale

In Ex. 5.2 the feature map values $F_{edge}(x,y)$ and $F_{blob}(x,y)$ are made semantically equivalent by converting them into evidence that a target is present at (x,y). In general, if we measure the evidence using a probabilistic scale, then mathematically we may represent the conversion of a feature map $F(x,y)$ into a probabilistic map $p(x,y)$ as follows

$$p(x,y) = S(F(x,y)|\alpha,\beta,\ldots,\gamma) ,$$

where S denotes a a parametric transfer function with parametes $\alpha,\beta,\ldots,\gamma$ [3].

The parameters $\alpha,\beta,\ldots,\gamma$ are unknown but may be learnt off-line as follows. Let $\mathbf{T} = (T_1,T_2,\ldots,T_M)^T$ denote a training set of M samples $T_m, m \in \{1,2,\ldots,M\}$. We suppose that each training sample T_m is characterized by a feature value F_m and an indicator function δ_m, where

$$\delta_m = \begin{cases} 1 \text{ if a target is associated with } T_m , \\ 0 \text{ otherwise} . \end{cases}$$

Let $\alpha^*, \beta^*, \ldots, \gamma^*$ denote the optimal (maximum likelihood) estimate of the parameters $\alpha, \beta, \ldots, \gamma$. Then $\alpha^*, \beta^*, \ldots, \gamma^*$ are learnt by maximizing the likelihood of F_m. Mathematically, we have

$$\begin{aligned}(\alpha^*, \beta^*, \ldots, \gamma^*) = \arg \max_{\alpha,\beta,\ldots,\gamma} \Big(\sum_{m=1}^{M} \delta_m \ln S(F_m|\alpha, \beta, \ldots, \gamma) \\ + (1 - \delta_m) \ln(1 - S(F_m|\alpha, \beta, \ldots, \gamma)) \Big) .\end{aligned} \tag{5.1}$$

Note: Eqn. (5.1) is equivalent to minimizing the sum of the square errors:

$$(\alpha^*, \beta^*, \ldots, \gamma^*) = \arg \min_{\alpha,\beta,\ldots,\gamma} \sum_{m=1}^{M} (S(F_m|\alpha, \beta, \ldots, \gamma) - \delta_m)^2 .$$

5.2.1 Plat Calibration

In Platt calibration [9] we assume the transfer function $S(F|\alpha, \beta, \ldots, \gamma)$ has a simple sigmoid shape with two parameters α and β:

$$S(F|\alpha, \beta) = \frac{1}{1 + \exp(\alpha(\beta - F))} .$$

In this case [7, 8, 9], the optimal (maximum likelihood) parameter estimates α^* and β^* are given by

$$(\alpha^*, \beta^*) = \arg\max_{\alpha,\beta} \Big(\sum_{m=1}^{M} \delta_m \ln S(F_m|\alpha, \beta) + (1 - \delta_m) \ln(1 - S(F_m|\alpha, \beta)) \Big) .$$

This optimization procedure is liable to overfit if the number of pixels with $\delta_m = 0$ or with $\delta_m = 1$ is too low. We may, however, mitigate the effects of over-fitting by replacing δ_m by a modified function δ'_m in the above minimization procedure, where

$$\delta'_m = \begin{cases} \dfrac{1}{M + 2 - \Delta} & \text{if } \delta_m = 0 , \\ \dfrac{1 + \Delta}{2 + \Delta} & \text{if } \delta_m = 1 , \end{cases}$$

and

$$\Delta = \sum_{m=1}^{M} \delta_m .$$

The following example describes the use of Platt calibration to semantically align several feature maps.

Example 5.3. Multi-Feature Infra-Red Target Detection in an Input Image [13]. We consider the detection of a small target in an infra-red input image I.

At each pixel (m,n) in I we test for the presence of a target by extracting the following local features:

1. Maximum Gray Level.
2. Contrast Mean Difference.
3. Average Gradient.
4. Gray-level Variation.
5. Entropy.

The five F_k clearly do not measure the same phenomena. However, according to the theory of target detection in an infra-red image, they are all causally linked to the presence of a target. Platt calibration is used to make the $F_k(m,n)$ semantically equivalent by converting each $F_k(m,n)$ into the probability, or likelihood, that a target is present at (m,n).

5.2.2 *Histogram Calibration*

In histogram calibration the feature space is divided into K non-overlapping bins $[t_{k-1},t_k), k \in \{1,2,\ldots,K\}$. In each bin, the transfer function is a flat plateau of constant height. In this case, the transfer function is given by:

$$S(F|\alpha,\beta,\ldots,\gamma) = \begin{cases} \alpha \text{ if } t_0 \leq F < t_1 \,, \\ \beta \text{ if } t_1 \leq F < t_2 \,, \\ \vdots \\ \gamma \text{ if } t_{K-1} \leq F < t_K \,, \end{cases}$$

and the optimal (maximum likelihood) estimates of the parameters $\alpha,\beta,\ldots,\gamma$ are:

$$\alpha^* = \frac{H_1}{M_1} \qquad \beta^* = \frac{H_2}{M_2} \qquad \ldots \qquad \gamma^* = \frac{H_K}{M_K} \,,$$

where M_k is the number of training samples which fall in the kth bin and H_k is the number of training samples which fall in the kth bin and which have $\delta_m = 1$.

5.2.3 *Isotonic Calibration*

In isotonic calibration [15] the feature space is divided into M bins, one for each feature value F_m. In each bin the transfer function is a flat plateau of constant height. No restriction is placed on the heights of the plateaus except they are isotonic, i. e. the heights of the plateaus are either monotonically increasing or monotonically decreasing.

Mathematically, the isotonic transfer function is given by

$$S(F|\alpha,\beta,\ldots,\gamma) = \begin{cases} \alpha \text{ if } t_0 \leq F < t_1 \ , \\ \beta \text{ if } t_1 \leq F < t_2 \ , \\ \vdots \\ \gamma \text{ if } t_{M-1} \leq F < t_M \ , \end{cases}$$

where $\alpha \leq \beta \leq \ldots \leq \gamma$ (monotonically increasing function) or $\alpha \geq \beta \geq \ldots \geq \gamma$ (monotonically decreasing function), $t_0 = -\infty$, $t_M = \infty$, and $t_m = (F_{m-1} + F_m)/2, m \in \{1,2,\ldots,M-1\}$.

The optimal (maximum likelihood) estimates of the parameters $\alpha,\beta,\ldots,\gamma$ are found by minimizing the sum of the errors in (5.1) assuming a monotonically varying transfer function F. A simple algorithm for doing this is the pooled-average value (PAV) algorithm [15].

Example 5.4. Platt, Histogram and Isotonic Calibration. In Table 5.1 we list a sequence of $M = 15$ training samples $T_m, m \in \{1,2,\ldots,M\}$, with feature values F_m and indicator functions δ_m. The corresponding Platt, histogram and isotonic calibration curves are shown in Fig. 5.1

Table 5.1 Isotonic Calibration

T_m	1	2	3	4	5	6	7	8	9	10	11	12	13	14	15
F_m	0.02	0.10	0.18	0.20	0.27	0.30	0.35	0.40	0.45	0.50	0.55	0.60	0.70	0.80	0.90
δ_m	0	1	0	0	1	0	1	1	0	1	1	1	0	1	1

5.3 Decision Labels

In decision fusion the decision maps D_k are obtained by performing a decision procedure on all pixels (x,y) in an input image I_k or in a feature map $F_k(x,y)$. For each pixel (x,y), $D_k(x,y)$ is a label l which may be any identifying name or symbol. We often find it convenient to associate each label l with an integer chosen from $l \in \{1,2,\ldots,L\}$. Let A and B be two decision maps with labels $m, m \in \{1,2,\ldots,M\}$, and $n, n \in \{1,2,\ldots,N\}$. Then, one way of making A and B semantically equivalent is to find which labels m in A are associated with labels n in B and vice versa. A convenient way of defining the associations is through an assignment matrix λ, where

$$\lambda(m,n) = \begin{cases} 1 \text{ if labels } m \text{ and } n \text{ are associated with each other} \ , \\ 0 \text{ otherwise.} \end{cases}$$

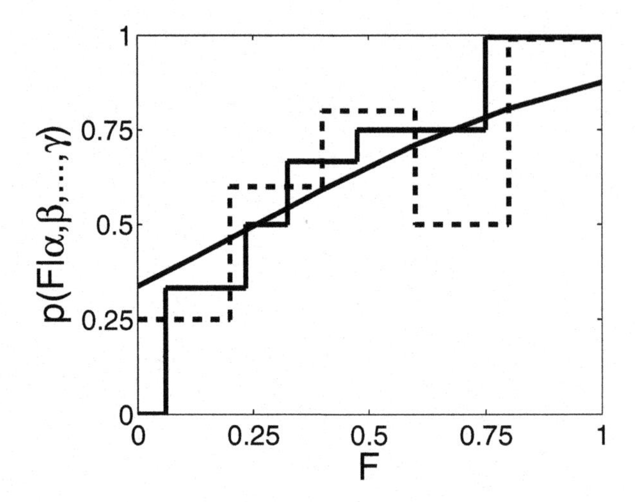

Fig. 5.1 Shows the Platt (continuous full curve), histogram (discontinuous dashed curve) and isotonic (discontinuous full curve) calibration curves obtained for the data listed in Table 5.1.

5.3.1 Assignment Matrix

If $C(m,n)$ denotes the cost of matching a label m with n, then the *optimal* assignment matrix $\widetilde{\lambda}$ is defined as the assignment matrix with the minimum overall cost:

$$\widetilde{\lambda} = \arg\min \sum_{m,n} C(m,n)\lambda(m,n) , \tag{5.2}$$

We often assume one-to-one associations between the labels m in A and the labels n in B. In this case we solve (5.2) subject to the following constraints:

$$\sum_{m=1}^{M} \lambda(m,n) \leq 1 , \qquad \sum_{n=1}^{N} \lambda(m,n) \leq 1 .$$

Fast algorithms for finding the optimal assignment matrix $\widetilde{\lambda}$ are available. Among them is the Hungarian algorithm [6] which is widely used in many applications. Traditionally the assignment algorithms are used when $M = N$ and we require that all labels in A are matched to a corresponding label in B and vice versa. The algorithms may, however, be used when $M \neq N$, or when we wish to make the assignment robust against outliers by finding the best $K, K \leq \min(M,N)$, associations [1]. In this case we use an enlarged cost matrix C_e:

$$C_e(m,n) = \begin{cases} C(m,n) & \text{if } m \in \{1,2,\ldots,M\}, n \in \{1,2,\ldots,N\} , \\ P & \text{otherwise} , \end{cases}$$

[1] This case is used as a guard against outliers.

where P is the cost, or penalty, of not associating a label in A with any label in B or of not associating a label in B with any label in A.

In some applications it may not be possible to define a penalty P. In this case, we use the following (sub-optimal) algorithm to find the sub-optimal label permutation (see Sect. 7.7.3):

Example 5.5. Simple Iterative Assignment Algorithm. Given a cost matrix $C(m,n), m \in \{1,2,\ldots,M\}, n \in \{1,2,\ldots,N\}$, we find the best K one-to-one associations, where $K \leq \min(M,N)$, as follows:

1. Find the association pair (m_1,n_1) with the smallest cost.
2. Find the association pair (m_2,n_2) with the second smallest cost, where $m_2 \neq m_1$ and $n_2 \neq n_1$.
3. Find the association pair (m_3,n_3) with the third smallest cost, where $m_3 \neq \{m_1,m_2\}$ and $n_3 \neq \{n_1,n_2\}$.
4. Continue this process until we have K pairs: $(m_1,n_1),(m_2,n_2),\ldots,(m_K,n_K)$.

In some applications we may impose additional constraints on (5.2). For example, when we compare two two-dimensional shapes or contours (Ex. 3.5) we may require the order of the points to be preserved [11].

In many applications, solving the assignment, or label correspondence, problem is the fusion algorithm itself. This is illustrated in the following example.

Example 5.6. Handwritten Character Recognition [1]. Shape matching is a powerful method for automatic handwritten character recognition. In this technique we match the contour of an unknown character with the contour of a known character. Suppose the two contours are labeled, respectively, as A and B, where A is sampled at M points z_m^A and B is sampled at N points z_n^B. For each pair of points (m,n) we let

$$C(m,n) = d(S_m, S_n) ,$$

where S_m and S_n are, respectively, the shape contexts (see Ex. 3.5) of the points z_m^A and z_n^B and $d(S_m,S_n)$ is an appropriate distance measure. If $\widetilde{\lambda}$ denotes the optimal one-to-one assignment matrix, then we may use

$$\widetilde{C} = \sum_{(m,n)} C(m,n)\widetilde{\lambda}(m,n) ,$$

as a similarity measure: the smaller $\widetilde{C}$ the more likely A and B are the same character.

In some applications, it may not be feasible, or even possible, to define an optimal assignment matrix $\widetilde{\lambda}$. In this case, we may make the decision maps semantically equivalent by converting them into co-associative matrices [4, 5].

5.3.2 Co-association Matrix

Given a decision map $D(m,n), m \in \{1,2,\ldots,M\}, n \in \{1,2,\ldots,N\}$, let $z_i = (m_i,n_i)$, $i \in \{1,2,\ldots,MN\}$. Then we define the co-association matrix [4, 5] as

$$A(i,j) = \begin{cases} 1 \text{ if } D(m_i,n_i) = D(m_j,n_j) \ , \\ 0 \text{ otherwise} \ . \end{cases}$$

We may interpret the co-association matrix as the evidence provided by D for the unknown true decision map.

The following example illustrates the formation of an average co-association matrix $\overline{A}$ from three co-association matrices A_1, A_2 and A_3.

Example 5.7. Co-association Matrix [14]. Given a one-dimensional image

$$I = (I_1, I_2, \ldots, I_7)^T \ ,$$

we segment it using three different cluster algorithms. The results are three decision maps:

$$D_1 = (1\ 1\ 2\ 2\ 2\ 3\ 3)^T \ , \qquad D_2 = (3\ 3\ 2\ 2\ 3\ 3\ 1)^T \ ,$$
$$D_3 = (2\ 3\ 2\ 2\ 1\ 1\ 1)^T \ .$$

The corresponding co-association matrices are:

$$A_1 = \begin{pmatrix} 1\,1\,0\,0\,0\,0\,0 \\ 1\,1\,0\,0\,0\,0\,0 \\ 0\,0\,1\,1\,1\,0\,0 \\ 0\,0\,1\,1\,1\,0\,0 \\ 0\,0\,1\,1\,1\,0\,0 \\ 0\,0\,0\,0\,0\,1\,1 \\ 0\,0\,0\,0\,0\,1\,1 \end{pmatrix} , \qquad A_2 = \begin{pmatrix} 1\,1\,0\,0\,1\,1\,0 \\ 1\,1\,0\,0\,1\,1\,0 \\ 0\,0\,1\,1\,0\,0\,0 \\ 0\,0\,1\,1\,0\,0\,0 \\ 1\,1\,0\,0\,1\,1\,0 \\ 1\,1\,0\,0\,1\,1\,0 \\ 0\,0\,0\,0\,0\,0\,1 \end{pmatrix} ,$$

$$A_3 = \begin{pmatrix} 1\,0\,1\,1\,0\,0\,0 \\ 0\,1\,0\,0\,0\,0\,0 \\ 1\,0\,1\,1\,0\,0\,0 \\ 1\,0\,1\,1\,0\,0\,0 \\ 0\,0\,0\,0\,1\,1\,1 \\ 0\,0\,0\,0\,1\,1\,1 \\ 0\,0\,0\,0\,1\,1\,1 \end{pmatrix} .$$

The mean co-association matrix is

$$\bar{A} = \frac{1}{K}\sum_{k=1}^{K} A_k = \begin{pmatrix} 1 & \frac{2}{3} & \frac{1}{3} & \frac{1}{3} & \frac{1}{3} & \frac{1}{3} & 0 \\ \frac{2}{3} & 1 & 0 & 0 & \frac{1}{3} & \frac{1}{3} & 0 \\ \frac{1}{3} & 0 & 1 & 1 & \frac{1}{3} & 0 & 0 \\ \frac{1}{3} & 0 & 1 & 1 & \frac{1}{3} & 0 & 0 \\ \frac{1}{3} & \frac{1}{3} & \frac{1}{3} & \frac{1}{3} & 1 & \frac{2}{3} & \frac{1}{3} \\ \frac{1}{3} & \frac{1}{3} & 0 & 0 & \frac{2}{3} & 1 & \frac{2}{3} \\ 0 & 0 & 0 & 0 & \frac{1}{3} & \frac{2}{3} & 1 \end{pmatrix},$$

which may, in turn, be regarded as a co-association matrix whose elements vary continuously between 0 and 1. The closer $\bar{A}(i,j)$ is to 1, the more the original decision maps put the elements I_i and I_j into the same cluster and the stronger the bond between the elements I_i and I_j, and vice versa.

5.4 Software

CLUSTERPACK. A matlab toolbox for cluster ensemble algorithms. Authors: A. Strehl and J. Ghosh [12].

GPAV. A matlab toolbox for isotonic regression. Authors: Oleg Burdakov, Anders Grimvall and Oleg Sysoev [2].

HUNGARIAN ALGORITHM FOR LINEAR ASSIGNMENT PROBLEM. A matlab routine for solving the linear assignment problem. Available from matlab central depository. Author: Yi Cao.

STPRTOOL. A statistical pattern recognition toolbox. Authors: Vojtech Franc and Vaclav Hlovac. The toolbox contains a file mlsigmoid.m which performs Platt calibration.

5.5 Further Reading

The calibration techniques discussed in Sect. 5.2 are not robust against outliers. For the modifications required when outliers are present see [10].

References

1. Belongie, S., Malik, J., Puzicha, J.: Shape matching and object recognition using shape contexts. IEEE Trans. Patt. Anal. Mach. Intell. 24, 509–522 (2002)
2. Burdakov, O., Grimvall, A., Sysoev, O.: Data preordering in generalized PAV algorithm for monotonic regression. J. Comp. Math. 24, 771–790 (2006)
3. Gebel, M., Weihs, C.: Calibrating classifier scores into probabilities. Adv. Data Anal., 141–148 (2007)

4. Fred, A.L.N.: Finding consistent clusters in data partitions. In: Kittler, J., Roli, F. (eds.) MCS 2001. LNCS, vol. 2096, pp. 309–318. Springer, Heidelberg (2001)
5. Fred, A.L.N., Jain, A.K.: Combining multiple clusterings using evidence accumulation. IEEE Trans. Patt. Anal. Mach. Intell. 27, 835–850 (2005)
6. Kuhn, H.W.: The Hungarian method for the assignment problem. Naval Research Logistics 52, 7–21 (2005)
7. Lin, H.T., Lin, C.J., Weng, R.: A note on Platt's probabilistic outputs for support vector machines. Mach. Learn. 68, 267–276 (2007)
8. Milgram, J., Cheriet, M., Sabourin, R.: Estimating accurate multi-class probabilities with support vector machines. In: Int. Joint Conf. Neural Networks (2005)
9. Platt, J.: Probabilistic outputs for support vector machines and comparisons to regularized likelihood methods. In: Smola, A.J., Bartlett, P., Scholkopf, B., Schurmans, D. (eds.) Advances in Large Margin Classifiers, pp. 61–74. MIT Press, Cambridge (1999)
10. Ruping, S.: Robust probabilistic calibration. In: Fürnkranz, J., Scheffer, T., Spiliopoulou, M. (eds.) ECML 2006. LNCS (LNAI), vol. 4212, pp. 743–750. Springer, Heidelberg (2006)
11. Scott, C., Nowak, R.: Robust contour matching via the order preserving assignment problem. IEEE Trans. Image Process. 15(9), 1831–1838 (2006)
12. Strehl, A., Ghosh, J.: Cluster ensembles - a knowledge reuse framework for combining multiple partitions. J. Mach. Learn. Res. 3, 583–617 (2002)
13. Wang, Z., Gao, C., Tian, J., Lia, J., Chen, X.: Multi-feature distance map based feature detection of small infra-red targets with small contrast in image sequences. In: Proc. SPIE, vol. 5985 (2005)
14. Wang, X., Yang, C., You, J.: Spectral aggregation for clustering ensemble. In: Proc. Int. Conf. Patt. Recogn., pp. 1–4 (2008)
15. Zadrozny, B., Elkan, C.: Transforming classifier scores into accurate multiclass probability estimates. In: Proc. Int. Conf. KDD (2002)

Chapter 6
Radiometric Calibration

Abstract. The subject of this chapter is radiometric calibration. This is the conversion of the input image values to a common radiometric scale. The transformation to such a scale is of critical importance in image fusion. Without a common radiometric base it is not possible to fuse images which were acquired at different illuminations, or under different atmospheric conditions or captured by different sensors. Radiometric calibration is used in both image fusion and in feature map fusion. For the sake of concreteness, we shall concentrate on the radiometric calibration of two input images A and B.

6.1 Introduction

The transformation to a common radiometric base is known as radiometric calibration, or normalization, which may in turn be divided into two types: absolute and relative. The absolute radiometric correction converts the digital counts of a pixel in the input image to radiance values. The absolute radiometric correction tends to be more accurate than the relative correction, but it needs sensor parameters, atmospheric refraction parameters and other data that are difficult to obtain. The difficulty in obtaining the above accurate atmospheric and sensor parameters makes relative radiometric normalization an attractive alternative. In relative radiometric normalization we designate one image as a reference image and adjust the radiometric properties of the second, or floating, image to match the reference image. The normalized image should therefore appear to have been acquired under the same conditions as the reference image.

Example 6.1. Intensity Standardization in MR Images [12]. Brain MR images present significant variations across patients and scanners. Consequently, training a classifier on a set of images and subsequently using it for brain segmentation may yield poor results. Significantly better segmentation is obtained if the image intensities are standardized beforehand.

6.2 Histogram Matching

In this section we consider the radiometric calibration technique known as histogram matching. This is one of the simplest and effective unsupervised radiometric alignment techniques. It uses the equalization of the histogram of the image B to the histogram of the reference image A. The method is a statistical method and does not require the accurate spatial alignment of the two images. It is therefore useful for calibrating images of the same scene which are acquired on different dates or with different illumination or atmospheric effects.

The goal of histogram equalization is to transform the image B in such a way that its pdf matches the pdf of the reference image A. Suppose b denotes a given pixel feature whose probability density function (pdf) and cumulative distribution function (cdf) are, respectively, $p_B(b)$ and $c_B(b)$, where $c_B(b) = \int_{-\infty}^{b} p_B(b)db$. We seek a function $a = F(b)$ which maps $p_B(b)$ into the corresponding reference pdf, $p_A(a)$. This is obtained by equating $c_B(b)$ and $c_A(a)$, where $c_A(a) = \int_{-\infty}^{a} p_A(a)da$:

$$c_B(b) = c_A(a) = c_A(F(b)) ,$$

or

$$a = F(b) = c_A^{-1}(c_B(b)) ,$$

where c_A^{-1} denotes the inverse of c_A [1].

If the pixel gray-levels in B are distinct, then histogram matching B to A is straightforward as the following example shows.

Example 6.2. Histogram Matching. Let A and B be two discrete images. Each image has M pixels. Let $H = (H_1, H_2, \ldots, H_L)$ be the histogram of A where H_l is the number of pixels in A with gray-level G_l. Let B^* denote the image B after histogram matching. If the pixel gray-levels in B are all distinct, then each pixel has a unique rank r_m associated with it. Then the histogram matching procedure is as follows.

```
R2 = 0
for l = 1:L
    R1 = R2 + 1;  R2 = R1 + Hl;
    for m = 1:M
      if (R1 <= rm <= R2);  B*m = Gl;  end
    end
end
```

[1] c^{-1} is defined as follows: If $y = c(x)$, then $c^{-1}(y) = x$.

6.2.1 Exact Histogram Specification

In many cases the number of pixels in an image, or an image patch, is much larger than the number of gray-levels. In this case, in order to obtain an exact histogram matching we require a method for ordering all the pixels which have the same gray-level. Traditionally, we order the pixels randomly. A better alternative is the following [2]. Separately convolve the image B with K small convolution masks $M_k, k \in \{1,2,\ldots,K\}$. Colute *et al.* [2] recommends the following six masks:

$$M_1 = \begin{pmatrix} 0\,0\,0\,0\,0 \\ 0\,0\,0\,0\,0 \\ 0\,0\,1\,0\,0 \\ 0\,0\,0\,0\,0 \\ 0\,0\,0\,0\,0 \end{pmatrix}, \quad M_2 = \begin{pmatrix} 0\,0\,0\,0\,0 \\ 0\,0\,1\,0\,0 \\ 0\,1\,1\,1\,0 \\ 0\,0\,1\,0\,0 \\ 0\,0\,0\,0\,0 \end{pmatrix}, \quad M_3 = \begin{pmatrix} 0\,0\,0\,0\,0 \\ 0\,1\,1\,1\,0 \\ 0\,1\,1\,1\,0 \\ 0\,1\,1\,1\,0 \\ 0\,0\,0\,0\,0 \end{pmatrix},$$

$$M_4 = \begin{pmatrix} 0\,0\,1\,0\,0 \\ 0\,1\,1\,1\,0 \\ 1\,1\,1\,1\,1 \\ 0\,1\,1\,1\,0 \\ 0\,0\,1\,0\,0 \end{pmatrix}, \quad M_5 = \begin{pmatrix} 0\,1\,1\,1\,0 \\ 1\,1\,1\,1\,1 \\ 1\,1\,1\,1\,1 \\ 1\,1\,1\,1\,1 \\ 0\,1\,1\,1\,0 \end{pmatrix}, \quad M_6 = \begin{pmatrix} 1\,1\,1\,1\,1 \\ 1\,1\,1\,1\,1 \\ 1\,1\,1\,1\,1 \\ 1\,1\,1\,1\,1 \\ 1\,1\,1\,1\,1 \end{pmatrix}.$$

Let $B_k(x,y), k \in \{1,2,\ldots,K\}$, denote the K outputs at the pixel (x,y), where by definition, $B_1(x,y) = B(x,y)$. We then order the pixels using the $B_k(x,y)$ as follows:

Example 6.3. Exact Histogram Specification [2].

```
for k=1:K
   If no ties exist, stop.
   Otherwise attempt to resolve ties using B_k.
end
If ties still exist resolve them randomly.
```

Once we have uniquely ordered the pixels according to b, i. e. each pixel has a unique integer rank $r(x,y)$ associated with it, we may then implement an exact histogram match as described in Ex. 6.2. *Note.* On the basis of r we may define a new image B', where

$$B'(x,y) = B(x,y) + \alpha r(x,y) ,$$

and α is a very small number [2]. By definition, the pixel gray-levels $B'(x,y)$ are unique but are still very close to the original gray-levels $B(x,y)$.

[2] If Δ denotes the smallest distance between adjacent gray-levels, then α should be less than Δ/N, where N is the number of pixels in the image.

Fig. 6.1 illustrates histogram equalization [3] using the exact histogram algorithm.

Fig. 6.1 **(a)** Shows an input image I with a full range of gray-levels. **(b)** Shows the result I_5 of histogram equalizing I to 5 levels. **(c)** Shows the result I'_4 of histogram equalizing I_5 to 4 levels using the traditional algorithm in which ties are randomly broken. **(d)** Shows the result I''_4 of histogram equalizing I_5 into 4 levels in which ties are broken using the exact histogram matching technique. We clearly see the improvement in image quality between I'_4 and I''_4.

6.3 Midway Image Equalization

Midway image equalization [3, 4] is defined as any method which warps two input histograms $p_A(a)$ and $p_B(b)$ to a common "intermediate" histogram p_Z, such that $p_Z(z)$ retains as much as possible of the shapes of $p_A(a)$ and $p_B(b)$. Mathematically,

[3] Histogram equalization is defined as a transformation of the gray-levels of an input image such that all gray-levels are equally populated.

[4] defines midway image equalization as follows: Given the two cumulative probability distributions $c_A(a)$ and $c_B(b)$ we define the intermediate distribution $p_Z(z)$ as the distribution whose inverse cumulative distance $c_Z^{-1}(z)$ is:

$$c_Z^{-1}(z) = \frac{c_A^{-1}(z) + c_B^{-1}(z)}{2} .$$

Suppose we warp $p_A(a)$ by matching it to $p_Z(z)$. Let $c_Z(z) = \int_0^z p_Z(z)dz$ denote the cumulative distribution of $p_Z(z)$, then the warped distribution is $p'_A(a')$, where

$$p'_A(a') = p_A(a) ,$$
$$a' = c_Z^{-1}(c_A(a)) .$$

Similarly we warp $p_B(b)$ by matching it to $p_Z(z)$. The corresponding warped distribution is $p'_B(b')$, where

$$p'_B(b') = p_Z(b) ,$$
$$b' = c_Z^{-1}(c_B(b)) .$$

The midway image equalization procedure may be implemented in an efficient manner using a dynamic programming technique [3]. However, when the images are of the same size and pixel gray-levels in each image are unique (i. e. no two pixels have the same gray-level) [4] we may implement a midway image equalization scheme as explained in the following example.

Example 6.4. Simple Midway Image Equalization. Given two $M \times N$ input images A and B in which all pixels have a unique gray-level, let $a_{(i)}$ and $b_{(i)}$ denote, respectively, the ith smallest gray-level in A and B. Then the corresponding midway gray-level is c_i, where

$$c_i = (a_{(i)} + b_{(i)})/2 .$$

Pseudo-code for calculating the midway equalized images A' and B' is:

```
[As,invrA] = sort(A(:)); [junk,rA] = sort(invrA);
[Bs,invrB] = sort(B(:)); [junk,rB] = sort(invrB);
C = (As+Bs)/2;
for i = 1 : M*N
    h = rA(i); A'(i) = C(h); k = rB(i); B'(i) = C(k);
end
A' = reshape(A',M,N); B' = reshape(B',M,N);
```

[4] If the pixels in the images are not unique then we may use the exact histogram specification scheme (Sect. 6.2.1) to create images whose gray-levels are unique.

6.4 Matching Second-Order Statistics

A simple version of histogram matching is to match the second-order statistics of the input images. Given two input images A and B of the same scene, we map the gray-levels of B so that the mean and standard deviation of B matches that of A. Let μ_A and μ_B denote the mean gray-level of A and B and let σ_A and σ_B denote the standard deviation of A and B. If $\widetilde{B}(m,n)$ denotes the gray-level of B after scaling, then

$$\widetilde{B}(m,n) = (B(m,n) - \mu_B)\frac{\sigma_A}{\sigma_B} + \mu_A .$$

Although very simple, matching second-order statistics is still widely used in remote sensing applications. In fact, in some applications [7], it may be preferred over histogram matching. For example, in merging infra-red and visible light images, Li and Wang [7] found it is preferable to perform radiometric calibration by matching the second-order statistics of the two images.

6.5 Ranking

Ranking is a robust method for radiometric normalization which like histogram matching does not require any training data. The following example illustrates the concept of ranking in remote sensing.

Example 6.5. Remote Sensing [8]. Remotely sensed data are increasingly used for mapping and monitoring the physical environment. One of the advantages of monitoring with remotely sensed data is that temporal sequences can accurately indicate environmental changes, assuming that the input data is radiometrically consistent for all scenes. Factors contributing to the potential inconsistency in measured radiance include changes in surface condition, illumination geometry, sensor calibration, observation geometry and atmospheric condition. By using a radiometric normalization technique, we may however, correct for data inconsistencies resulting from many different effects. Image normalization is carried out in one step by converting image values to ordinal ranks. Ordinal ranking allows us to assign each pixel a new value based on its reflectance value, relative to all other pixels. When image pairs are converted to ordinal ranks the global characteristics of the distributions of pixel values are matched.

Pixel ranking does not require atmospheric details, sensor information, or selection of subjective pseudo-invariant features, and therefore allows images to be simply and efficiently normalized and processed for changes with minimal *a priori* knowledge. In general, for small pictures, pixel ranking is an effective image normalization technique. It is less effective on very large digital images because in this case we obtain many tied ranks, although the exact ordering technique discussed in Sect. 6.2.1 may help.

6.6 Thresholding

In image thresholding we convert an input image I into a binary labeled image B using a threshold gray-level t:

$$B(x,y) = \begin{cases} 1 \text{ if } I(x,y) \geq t \,, \\ 0 \text{ otherwise} \,. \end{cases} \tag{6.1}$$

The primary purpose of a thresholding algorithm is to segment the input image into background regions and foreground regions or objects of interest. However, image thresholding is sometimes also used as a simple method for radiometric normalization. The following algorithm describes the Otsu thresholding algorithm.

Example 6.6. Otsu thresholding algorithm [9]. Given an input picture I, let $g(x,y)$ denote the gray-level at pixel (x,y). Then pixel gray-levels are divided into two groups: foreground pixels whose gray-levels are less than, or equal to, a threshold t and background pixels whose gray-levels are greater than t. The optimum threshold is found by maximizing the separation between the two groups. Let $\mu_F(t)$, $\sigma_F(t)$ and $\mu_B(t)$, $\sigma_B(t)$ denote, respectively, the mean gray-level and standard deviation of the foreground and background pixels (defined with a threshold t), then the optimum threshold is given by

$$t_{OPT} = \arg\max_t \left(\frac{P(t)\,(1-P(t))\,(\mu_F(t)-\mu_B(t))^2}{P(t)\sigma_F^2(t)+(1-P(t))\,\sigma_B^2(t)} \right) ,$$

where $P(t)$ is the relative number of pixels with gray-level less than, or equal to, t.

In some applications the conversion of the pixel gray-levels $I(x,y)$ in the input image into binary gray-levels $B(x,y)$ using (6.1) is too coarse. In these cases, we may use a fuzzy thresholding algorithm which generates a fuzzy gray-level image $\widetilde{B}$, where $\widetilde{B}(x,y) \in [0,1]$.

Fig. 6.2 shows the result of thresholding an input image.

The following example illustrates the use of local thresholding for radiometric calibration.

Example 6.7. Local Binary Pattern for Radiometric Calibration [5, 6]. Ref. [5, 6] describes the use of the local binary pattern (LBP) operator as an efficient method for radiometric calibration of face images in an *uncontrolled* environment. The LBP operator (see Sect. 3.4) works as follows. It takes a local neighborhood around each pixel and thresholds the pixels in the neighborhood according to the value of the center pixel. The weighted sum of the thresholded pixels is a label which may be regarded as a radiometrically calibrated pixel value (see Fig. 3.5). For a 3×3 neighborhood centered on the

(a)

(b)

Fig. 6.2 **(a)** Shows an input image I. **(b)** Shows the binary image after thresholding I with the Otsu algorithm.

pixel (m,n), the LBP operator is defined as

$$LBP(m,n) = \sum_{k=0}^{7} s(A(m,n),A(i_k,j_k))2^{(k-1)} ,$$

where (i_k, j_k) are the coordinates of the kth pixel in the 3×3 neighborhood of (m,n) and

$$s(A(m,n),A(i_k,j_k)) = \begin{cases} 1 \text{ if } A(i_k,j_k) > A(m,n) , \\ 0 \text{ otherwise} . \end{cases}$$

Note: The operator may be extended to circular neighborhoods by bilinearly interpolating the pixel values [6].

6.7 Segmentation

In image segmentation we convert an input image into a multiple label image. Although segmentation is primarily a diagnostic tool in which the input image is decomposed into contiguous regions, we may also use it as a method for radiometric normalization. The following example illustrates a simple K-means cluster algorithm which may be used for image segmentation.

Example 6.8. K-means cluster algorithm. Given an input image I with pixel gray-levels $g_m, m \in \{1,2,\ldots,M\}$. Let $G_1, G_2, \ldots, G_K$ denote K cluster centers or cluster gray-levels. Each pixel gray-level g_m is associated with a given cluster:

$$\delta_{mk} = \begin{cases} 1 \text{ if } g_m \text{ is associated with } G_k , \\ 0 \text{ otherwise} . \end{cases}$$

Then the K-means algorithm attempts to find the set of cluster centers $G_k, k \in \{1,2,\ldots,K\}$, such that the total error is a minimum:

$$(G_1, G_2, \ldots, G_K) = \arg\min_{G_k} \sum_{m=1}^{M} \sum_{k=1}^{K} \delta_{mk} C(g_m, G_k) ,$$

where C is an appropriate cost function. A common cost function is $C(x,y) = |x-y|$.

The K-means algorithm works in an iterative manner as follows: In each iteration we calculate the assignment matrix δ_{mk} using the cluster centers G_k calculated in the previous iteration. The cluster centers are then recalculated using the new assignment matrix. The process for T iterations is:

```
for t = 1:T
   for m = 1:M
     δ_mk^(t) = { 1 if |g_m − G_k^(t−1)| = min_l |g_m − G_l^(t−1)|;
                { 0 otherwise;
   end
   for k = 1:K
     G_k^(t) = Σ_{m=1}^M δ_mk^(t) g_m / Σ_{m=1}^M δ_mk^(t);
   end
end
```

Fig. 6.3 illustrates the segmentation of an input image using the K-means cluster algorithm. Although very simple the K-means cluster algorithm is widely used as a method of image segmentation. Recently with the development of ensemble learning, the K-means algorithm is found to be capable of giving state-of-the-art segmentation (see Chapt. 16).

Note: Although image segmentation is an effective method for radiometric calibration, the segmented images may still require semantic equalization (see Sect. 5.3).

6.8 Feature Map Normalization

Although in many cases, feature map normalization requires the maps to be brought into semantic equivalence, there are cases when the feature maps measure the same object, or phenomena, and semantic equivalence is not required. In this case, the feature maps may be normalized using any of the techniques discussed previously.

(a)

(b)

Fig. 6.3 **(a)** Shows an input image I. **(b)** Show the segmentation of I using the K-means algorithm with $K = 6$.

Example 6.9. Multiple Edge Maps. We consider an input image on which we apply a Sobel and a Canny edge detectors. The two detectors work on different principles but both measure the presence, or otherwise, of an edge in the input image. The two feature maps, F_{sobel} and F_{canny}, are clearly semantically equivalent. They may therefore be fused together if the feature maps F_{sobel} and F_{canny} are radiometrically aligned to the same scale. If we use a simple linear radiometric scale, the corresponding calibrated maps are:

$$\lambda_{sobel}(x,y) = \frac{F_{sobel}(x,y) - F_{sobel}^{\min}}{F_{sobel}^{\max} - F_{sobel}^{\min}} ,$$

$$\lambda_{canny}(x,y) = \frac{F_{canny}(x,y) - F_{canny}^{\min}}{F_{canny}^{\max} - F_{canny}^{\min}} ,$$

where $F_{sobel}^{\min} = \min_{(x,y)} F_{sobel}(x,y)$, $F_{sobel}^{\max} = \max_{(x,y)} F_{sobel}(x,y)$, $F_{canny}^{\min} = \min_{(x,y)} F_{canny}(x,y)$ and $F_{canny}^{\max} = \max_{(x,y)} F_{canny}(x,y)$.

6.9 Probabilistic Scale

In Ex. 5.2 we described making two feature maps semantically equivalent by converting them into probabilistic, or likelihood, maps. This transformation may also be used for radiometric calibration.

6.10 Software

GPAV. A matlab toolbox for isotonic regression. Authors: Oleg Burdakov, Anders Grimvall and Oleg Sysoev [1].

LIBRA. A toolbox for performing classical and robust statistics. The toolbox contains m-files on various robust normalization techniques. Authors: Sabine Verboven and Mia Hubert [11].

MATLAB STATISTICAL TOOLBOX. Matlab statistical toolbox. The toolbox contains m-files for performing various radiometric calibration procedures.

STPRTOOL. A statistical pattern recognition toolbox. Authors: Vojtech Franc and Vaclav Hlovac. The toolbox contains a file mlsigmoid.m which performs Platt calibration.

6.11 Further Reading

In this chapter we have given a brief overview of some relative calibration methods which have general applicability. However, for specific applications, specialized normalization techniques may be available. For example, [10] contains a comparison of different radiometric calibration algorithms for face verification.

References

1. Burdakov, O., Grimvall, A., Sysoev, O.: Data preordering in generalized PAV algorithm for monotonic regression. J. Comp. Math. 24, 771–790 (2006)
2. Coltue, D., Bolon, P., Chassery, J.-M.: Exact histogram specification. IEEE Trans. Image Process. 15, 1143–1152 (2006)
3. Cox, I., Roy, S., Hingorani, S.L.: Dynamic histogram warping of image pairs for constant image brightness. In: Proc. IEEE Int. Conf. Image Process., vol. 2, pp. 366–369 (1995)
4. Delon, J.: Midway image equalization. J. Math. Imag. Vis. 21, 119–134 (2004)
5. Heusch, G., Rodriguez, Y., Marcel, S.: Local binary patterns as an image pre-processing for face authentification. In: Proc. FGR (2006)
6. Holappa, J., Ahonen, T., Pietikainen, M.: An optimized illumination normalized method for face recognition. In: Second IEEE Int. Conf. Biometrics: Theory Appl. Systems (2008)
7. Li, G., Wang, K.: Merging infrared and color visible images with a contrast enhanced fusion method. In: Proc. SPIE, vol. 6571, p. 657108 (2007)
8. Nelson, T., Wilson, H.G., Boots, B., Wulder, M.A.: Use of ordinal conversion for radiometric normalization and change detection. Int. J. Remote Sensing 26, 535–541 (2005)
9. Otsu, N.: A threshold selection method from gray-level histogram. IEEE Trans. Syst. Man Cybernetics 9, 62–66 (1979)
10. Short, J., Kittler, J., Messer, K.: A comparison of photometric normalization algorithms for face authentification. In: Proc. AFGR (2004)
11. Verboven, S., Hubert, M.: Libra: A matlab library for robust analysis. Chemometrics and Intell. Laboratory Syst. 75, 127–136 (2005)
12. Zhuge, Y., Udupa, J.K.: Intensity standardization simplifies brain MR image segmentation. Comp. Vis. Image Under. 113, 1095–1103 (2009)

Chapter 7
Pixel Fusion

Abstract. The subject of this chapter is image fusion techniques which rely on simple pixel-by-pixel operations. The techniques include the basic arithmetic operations, logic operations and probabilistic operations as well as slightly more complicated mathematical operations. The image values include pixel gray-levels, feature map values and decision map labels. Although more sophisticated techniques are available, the simple pixel operations are still widely used in many image fusion applications.

7.1 Introduction

In this chapter we consider fusion techniques which rely on simple pixel operations on the input image values. We assume the input images are spatially and temporally aligned, semantically equivalent and radiometrically calibrated. We start with the image fusion of K input images $I_1, I_2, \ldots, I_K$ using a simple arithmetic addition operator.

7.2 Addition

Addition which is probably the simplest fusion operation. It works by estimating the average intensity value of the input images $I_k, k \in \{1, 2, \ldots, K\}$, on a pixel-by-pixel basis. If $\widetilde{I}(m,n)$ denotes the fused image at the pixel (m,n), then

$$\widetilde{I}(m,n) = \frac{1}{K} \sum_{k=1}^{K} I_k(m,n) . \tag{7.1}$$

Although extremely simple, (7.1) is widely used if the input images are of the same modality.

The technique assumes semantic alignment and requires very accurate spatial and radiometric alignment. The technique has the advantage of suppressing any noise which is present in the input images. The following example illustrates how the pixel addition technique reduces image noise in a video sequence.

Example 7.1. Video Noise Averaging [3]. We consider an efficient method for video denoising. Although we can apply static image denoising methods to the case of image sequences we can do much better by including temporal information (inter-frame methods). This temporal information is crucial since our perception is very sensitive to temporal distortions like edge displacement: the disregard of temporal information may lead to inconsistencies in the result.

The input to the denoising algorithm is a video sequence of $M \times N$ images $I_k, k \in \{1,2,\ldots\}$. We partition each image I_k into a disjoint set of horizontal lines $L_k^{(i)}$. For each line $L_k^{(i)}$ we consider the family of lines which are close to $L_k^{(i)}$ in the same image and in the neighouring images. We warp each of these lines so they match with $L_k^{(i)}$. Let $\phi(L_l^{(j)})$ denote the warped version of the line $L_l^{(j)}$ onto the line $L_k^{(i)}$. We then obtain a denoised version of $L_k^{(i)}$ by performing an average of the lines $\phi(L_l^{(j)})$.

The pixel average technique has the disadvantage that it tends to suppress salient image features producing a low contrast image with a "washed-out" appearance. This effect can be alleviated, to some extent, by using a linear weighted average of the input images:

$$\widetilde{I}(m,n) = \sum_{k=1}^{K} w_k I_k(m,n) \Big/ \sum_{k=1}^{K} w_k \,, \tag{7.2}$$

where w_k are pre-selected scalars which are chosen so that each input image contributes an "optimal" amount towards the fused image. For instance, when fusing thermal and electro-optical sensors we may assign larger weights to the warmer or the cooler pixels of the thermal image or we may assign larger weights to those pixels whose intensities are much different from its neighbors. In some applications we estimate the weights w_k using the expectation-maximization (EM) algorithm (see Ex. 7.2).

Instead of pre-selecting the weights w_k we may allow the weights to vary automatically according to the amount of information contained in I_k. One method of defining the weights w_k is to use the method of principal component analysis (PCA) (Sect. 9.2).

However, notwithstanding how the weights are chosen, pixel averaging will tend to reduce the contrast of an object if in one image the object appears with a certain contrast and in another image the object appears with the opposite contrast.

7.2.1 Robust Averaging

Instead of using the arithmetic mean we may use robust equivalents which are robust against outliers. Two such operators are the median operator and the trimmed mean operator:

Median operator

$$\tilde{I}(x,y) = \text{med}_k\,(I_k(x,y)) \ .$$

Trimmed mean operator

$$\tilde{I}(x,y) = \frac{1}{K-2\alpha} \sum_{k=\alpha+1}^{K-\alpha} I_{(k)}(x,y) \ ,$$

where $I_{(k)}(x,y) = I_l(x,y)$ if $I_l(x,y)$ is the lth largest gray-level at (x,y) and α is a small constant. We often set $\alpha = \lfloor K/20 \rfloor$.

7.3 Subtraction

Subtraction is the complement to addition and is used as a simple fusion operator in change detection algorithms. These algorithms apply the subtraction operator pixel-by-pixel to generate a signed difference image D:

$$D(x,y) = I_1(x,y) - I_2(x,y) \ ,$$

where I_1 and I_2 are two input images which have been carefully aligned. The difference image is then thresholded to create a change map $B(x,y)$, where

$$B(x,y) = \begin{cases} 1 \text{ if } |D(x,y)| > t \ , \\ 0 \text{ otherwise} \ . \end{cases}$$

The threshold t may be constant over the image D or it may vary from pixel to pixel. The following example illustrates the Bayesian approach to change detection

Example 7.2. Unsupervised Change Detection [2, 4]. Given a difference image we write it as a one-dimensional vector $\mathbf{D} = (D(1), D(2), \ldots, D(M))^T$. We assume the probability density of the difference values, $P(D)$, can be modeled as a mixture of $K = 2$ components: one component corresponding to the class c_1 of "change" pixels and the other component corresponding to the class c_2 of "no-change" pixels:

$$P(|D) = P(c_1)p(D|c_1) + P(c_2)p(D|c_2) ,$$
$$= \sum_{k=1}^{K} W_k p(c_k|D) ,$$

where $W_k = P(c_k)$ is the *a priori* probability of the class $c_k, k \in \{1,2\}$. The standard approach for finding the *a posteriori* probability $p(c_k|D(m))$ is the expectation-maximization (EM) algorithm. We assume the likelihood $p(D(m)|c_k)$ is Gaussian:

$$p(D(m)|c_k) = \frac{1}{\sigma_k\sqrt{2\pi}} \exp -\frac{1}{2}\left(\frac{D(m)-\mu_k}{\sigma_k}\right)^2 ,$$

where μ_k and σ_k are, respectively, the mean and standard deviation of the kth Gaussian distribution. Then the EM algorithm iteratively updates the *a posteriori* probability distribution $p(D(m)|c_k)$ that $D(m)$ was generated by the kth mixture component, the *a priori* class probabilities W_k, and the Gaussian parameters μ_k and σ_k. Each iteration t consists of two steps:

E-step. Update the *a posteriori* probability $p(D(m)|c_k)$:

$$p^{(t+1)}(c_k|D(m)) = W_k^{(t)} p(D(m)|\mu_k^{(t)}, \sigma_k^{(t)}) \Big/ \sum_{h=1}^{K} W_h^{(t)} p(D(m)|\mu_h^{(t)}, \sigma_h^{(t)}) .$$

M-step. Update the maximum likelihood estimates of the parameters $W_k^{(t)}, \mu_k^{(t)}$ and $\sigma_k^{(t)}$ for each component $k, k \in \{1,2,\ldots,K\}$:

$$W_k^{(t+1)} = \frac{1}{M} \sum_{m=1}^{M} p^{(t+1)}(c_k|D(m)) ,$$
$$\mu_k^{(t+1)} = \frac{\sum_{m=1}^{M} p^{(t+1)}(c_k|D(m))D(m)}{\sum_{m=1}^{M} p^{(t+1)}(c_k|D(m))} ,$$
$$(\sigma_k^2)^{(t+1)} = \frac{\sum_{m=1}^{M} p^{(t+1)}(c_k|D(m))(D(m)-\mu_k^{(t+1)})^2}{\sum_{m=1}^{M} p^{(t+1)}(c_k|D(m))} .$$

After several iterations the *a posteriori* probabilities $p^{(t)}(c_k|D(m))$ and the parameters $W_k^{(t)}$, $\mu_k^{(t)}$ and $\sigma_k^{(t)}$ converge to their final values. The $D(m)$ are then assigned to the class c_k with maximum *a posteriori* probability:

$$c_{opt} = \arg\max_k P(c_k|D(m)) ,$$
$$= \arg\max_k (P(c_k)p(D(m)|c_k)) .$$

We may generalize the above procedure by assuming generalized Gaussian likelihoods [2].

Example 7.3. Mixture of Generalized Gaussian Distributions [2]. Ref. [2] suggests that a better model for the likelihood $p(D(m)|c_k)$ is a generalized Gaussian distribution:

$$p(D(m)|\mu_k,\sigma_k,\alpha_k) = \frac{\lambda_1(\alpha_k)}{\sigma_k}\exp\left(-\lambda_2(\alpha_k)\left|\frac{D(m)-\mu_k}{\sigma_k}\right|^{\alpha_k}\right),$$

where

$$\lambda_1(\alpha_k) = \frac{\alpha_k\Gamma(3/\alpha_k)^{1/2}}{2\Gamma(1/\alpha_k)^{3/2}},$$

$$\lambda_2(\alpha_k) = \left(\frac{\Gamma(3/\alpha_k)}{\Gamma(1/\alpha_k}\right)^{\alpha_k/2}.$$

The advantage of using the generalized Gaussian distribution is that by changing α_k we may change the shape of $p(D(m)|\mu_k,\sigma_k,\alpha_k)$. For example, $p(D(m)|\mu_k,\sigma_k,\alpha_k)$ assumes, respectively, the form of an impulsive, Laplacian, Gaussian and uniform distribution as α_k adopts the values $0,1,2$ and ∞.

The EM algorithm [6] for the generalized Gaussian model is the same as the standard EM algorithm given above apart from an addition to the M-step, where we update the shape parameter α_k. In order to update α_k we first update the kurtosis of the distribution:

$$\kappa_k^{(t+1)} = \frac{\sum_{m=1}^{M} p^{(t)}(c_k|D(m))(D(m)-\mu_k^{(t+1)})^4}{(\sigma_k^{(t+1)})^4\sum_{m=1}^{M} p^{(t)}(c_k|D(m))} - 3,$$

and then calculate $\alpha_k^{(t+1)}$ using the following relationship:

$$\kappa_k^{(t+1)} = \frac{\Gamma(5/\alpha_k^{(t+1)})\Gamma(1/\alpha_k^{(t+1)})}{\Gamma(3/\alpha_k^{(t+1)})^2} - 3.$$

Apart from the EM algorithm, Chapt 12 contains a review of many formulas and algorithms used to threshold D.

The difference image is sensitive to noise and variations in illuminations. In general, therefore, difference images are only used if the input images were captured with the same sensor under similar conditions, i. e. the photometric transformation between corresponding pixel gray-levels values should be close to identity.

In the next two sections we consider the multiplication and division operators. In general, these operations are much less widely used than the addition and the subtraction operations.

7.4 Multiplication

Multiplication and division are not widely used as image fusion operators. However one important image fusion application where multiplication is used is Brovey pan-sharpening.

Example 7.4. Brovey Pan Sharpening [11]. The Brovey transform is a simple method for combining multi-spectral image with a panchromatic image. The technique is limited to three spectral bands which we identify with the R, G and B channels. The transform is defined as follows

$$\begin{pmatrix} R_{brovey} \\ G_{brovey} \\ B_{brovey} \end{pmatrix} = \begin{pmatrix} R \\ G \\ B \end{pmatrix} + (P-I) \begin{pmatrix} R/P \\ G/P \\ B/P \end{pmatrix} ,$$

where $I = (R+G+B)/3$ and P denotes the panchromatic image.

7.5 Division

The following example illustrates shadow detection by computing a ratio map R.

Example 7.5. Shadow Detection in Color Aerial Images [5, 10]. Shadow detection algorithms are very important in many image fusion algorithms. Generally these algorithms work by selecting a region which is darker than its neighboring regions but has similar chromatic properties. For *RGB* color aerial images we may detect shadows as follows. We transform the *RGB* input color image into a intensity-hue-saturation (*IHS*) color space (16.1–16.3). Then at each pixel (m,n) we form a ratio map $R_e(m,n)$ by comparing the hue of the pixel to the intensity of the pixel. The value $R_e(m,n)$ measures the likelihood of the pixel (m,n) being in shadow.

In [5] the ratio map is defined as follows (assuming 24-bit *RGB* input picture):

$$R_e(m,n) = \text{round}\left(\frac{H_e(m,n)}{I_e(m,n)+1}\right) ,$$

where

$$I_e(m,n) = \frac{R(m,n)+G(m,n)+B(m,n)}{3} ,$$

$$H_e(m,n) = \frac{255(\tan^{-1}(H(m,n)+\pi)}{2\pi} .$$

A shadow map is then formed by thresholding R_e:

$$S_e(m,n) = \begin{cases} 1 \text{ if } R_e(m,n) > T \ , \\ 0 \text{ otherwise} \ , \end{cases}$$

where $S_e(m,n) = 1$ denotes a shadow pixel at (m,n).

7.6 Feature Map Fusion

In feature map fusion we fuse together the feature maps $F_k, k \in \{1,2,\ldots,K\}$. The following example illustrates the fusion of multiple feature maps which are semantically equivalent. The fusion operator used is a simple arithmetic average operator applied separately to each pixel.

Example 7.6. Fusion of Multiple Edge Maps. Given an input image I we perform edge extraction using several edge operators e. g. Sobel, Canny and zero-crossing. The operators all measure the same phenomena ("presence of an edge") and are therefore semantically equivalent. The feature maps still require radiometric calibration onto a common scale. If $F_{sobel}(m,n)$, $F_{canny}(m,n)$ and $F_{zero}(m,n)$ denote the feature maps after calibration, then we may fuse the maps together using a simple arithmetic mean operator:

$$\widetilde{F}(m,n) = \frac{1}{3}(F_{sobel}(m,n) + F_{canny}(m,n) + F_{zero}(m,n)) \ .$$

The following example illustrates the fusion of multiple feature maps which do not measure the same phenomena but which have been made semantically equivalent by transforming them into probabilistic, or likelihood, maps.

Example 7.7. Multi-Feature Infra-red Target Detection in an Input Image [12]. We continue with Ex 5.3. We consider the detection of a small target in an infra-red input image I. At each pixel (m,n) in I we test for the presence of a target by extracting K features $F_k, k \in \{1,2,\ldots,K\}$. The features do not measure the same phenomena. However, according to the theory of infra-red target detection, they are all related to the presence of an infra-red target. We make the F_k semantically equivalent by transforming $F_k(m,n)$ into a probability $p_k(m,n)$ which measures the probability, or likelihood, of an infra-red target being present at (m,n). Let $\widetilde{p}(m,n)$ be the fused probability, or likelihood, that an infra-red target is present at (m,n). Then, methods for fusing the $p_k(m,n)$ include:

Mean

$$\widetilde{p}(m,n) = \frac{1}{K}\sum_{k=1}^{K} p_k(m,n) .$$

Product

$$\widetilde{p}(m,n) = \prod_{k=1}^{K} p_k(m,n) .$$

Minimum

$$\widetilde{p}(m,n) = \min_k p_k(m,n) .$$

Median

$$\widetilde{p}(m,n) = \text{median}(p_k(m,n)) .$$

Maximum

$$\widetilde{p}(m,n) = \max_k p_k(m,n) .$$

Another method for feature map fusion is rank fusion. The following example illustrates rank fusion for face recognition.

Example 7.8. Face Recognition Using Rank Fusion [7]. Given an input image I we extract several different features $F_k, k \in \{1,2,\ldots,K\}$. *Note*: In this example, the features F_k refer to the entire image I and not just to a pixel (x,y).

We make the F_k semantically equivalent by transforming each F_k into a multiple set of L likelihoods $p_k(l), l \in \{1,2,\ldots,L\}$, where $p_k(l)$ is the probability that the feature F_k belongs to the lth individual (i. e. belongs to class l). If the $p_k(l)$ are reliable we may fuse them together using any of the operators discussed in Ex. 7.7. However, in many cases we can only rely on the rank of $p_k(l)$ and not on the actual value of $p_k(l)$. In this case we transfer each $p_k(l)$ into a rank $r_k(l)$, where

$$r_k(l) = r \qquad \text{if } p_k(l) \text{ is } r\text{th largest likelihood} .$$

The optimum classification of the input image I is then

$$l^* = \arg\min_l(\widetilde{r}(l)) = \arg\min_l\Big(\sum_{k=1}^{K} r_k(l)\Big) , \tag{7.3}$$

where $\widetilde{r}$ denotes the sum of the ranks for class l:

$$\widetilde{r}(l) = \sum_{k=1}^{K} r_k(l) .$$

A simple numerical example illustrating the technique is as follows: We have three features: face matching (F_1), ear matching (F_2) and signal matching (F_3). The ranks obtained for each feature are:

$$\begin{array}{llll} r_1(1) = 3, & r_1(2) = 1, & r_1(3) = 4, & r_1(4) = 2\,, \\ r_2(1) = 2, & r_2(2) = 1, & r_2(3) = 4, & r_2(4) = 5\,, \\ r_3(1) = 1, & r_3(2) = 2, & r_3(3) = 3, & r_3(4) = 4\,. \end{array}$$

The fused ranks are:

$$\tilde{r}_1 = 6, \quad \tilde{r}(2) = 4, \quad \tilde{r}(3) = 11, \quad \tilde{r}(4) = 11\,.$$

The optimal classification is $l^* = \arg\min_l(\tilde{r}(l)) = \arg\min(6,4,11,11) = 2$.

7.7 Decision Fusion

In decision fusion we fuse together a set of decision images, or label maps, $D_k, k \in \{1,2,\ldots,K\}$. The D_k are themselves obtained by performing a decision procedure on all pixels (m,n) in the input image I_k. For each pixel (m,n), $D_k(m,n)$ is a label l which may be any identifying name or symbol. We shall find it convenient to associate each label l with an integer chosen from $l \in \{1,2,\ldots,L\}$. It should, however, be emphasized, that in general different labels have different meaning and this must be taken into account when the D_k are fused together.

We shall start by considering the case when the $D_k, k \in \{1,2,\ldots,K\}$, are semantically equivalent, i. e. a label l in D_h has the same semantic interpretation as the label l in $D_k, h \neq k$. Then in Sects. 7.7.3 and 7.7.4 we consider the more complicated case, when the D_k are no longer semantically equivalent.

The simplest way of fusing D_k which are semantically equivalent is to fuse the D_k using the majority-vote rule:

$$\tilde{D}(m,n) = l \quad \text{if} \quad \sum_{k=1}^{K} \delta\left(D_k(m,n),l\right) \geq \frac{1}{2}\,,$$

or the weighted majority-vote rule:

$$\tilde{D}(m,n) = l \quad \text{if} \quad \sum_{k=1}^{K} w_k \delta\left(D_k(m,n),l\right) \geq \sum_{k=1}^{K} w_k/2\,, \tag{7.4}$$

where

$$\delta(a,b) = \begin{cases} 1 \text{ if } a = b\,, \\ 0 \text{ otherwise}\,. \end{cases}$$

In some applications the weights in (7.4) may be obtained from the D_k themselves (see e. g. Ex. 10.7). In other cases we may obtain the weights using the expectation-maximization (EM) algorithm (see Ex. 7.2 and Chapt. 21).

The majority-vote and weighted majority-vote rules are widely used for decision fusion. They are simple to implement and robust against noise and outliers (see Fig. 7.1).

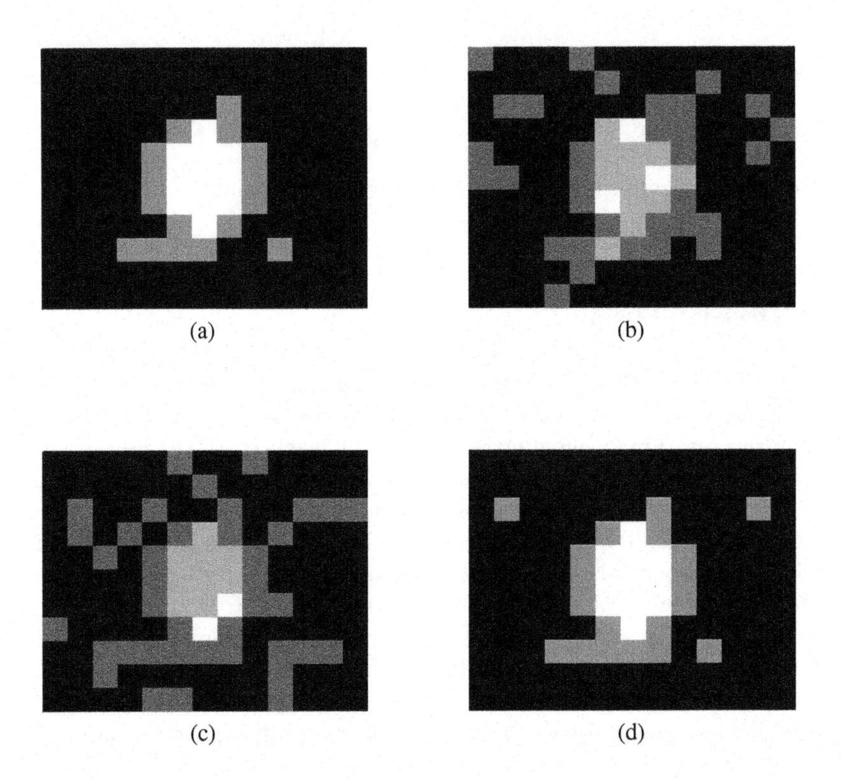

Fig. 7.1 (**a**) Shows a decision map D_1 in which the structures are contiguous. (**b**) and (**c**) Show decision maps D_2 and D_3. These are the same as D_1 and with additive noise. (**d**) Shows the decision map $\tilde{D}$ obtained by majority-vote fusion of D_1, D_2 and D_3. In this case, the fused map maintains the contiguous nature of the original input maps.

The majority-vote and the weighted majority-vote rules do not, however, take into account pixel-to-pixel correlations. In some cases this may lead to a fragmentation of structures which are contiguous in the input images (see Fig. 7.2). To prevent the fragmentation we must include the effect of pixel-to-pixel correlations. One way of doing this is to use a Markov random field which is discussed in Chapt. 17. Alternatively, if the fragmentation arises because the D_k are not perfectly aligned, then we may use the shape-based averaging algorithm.

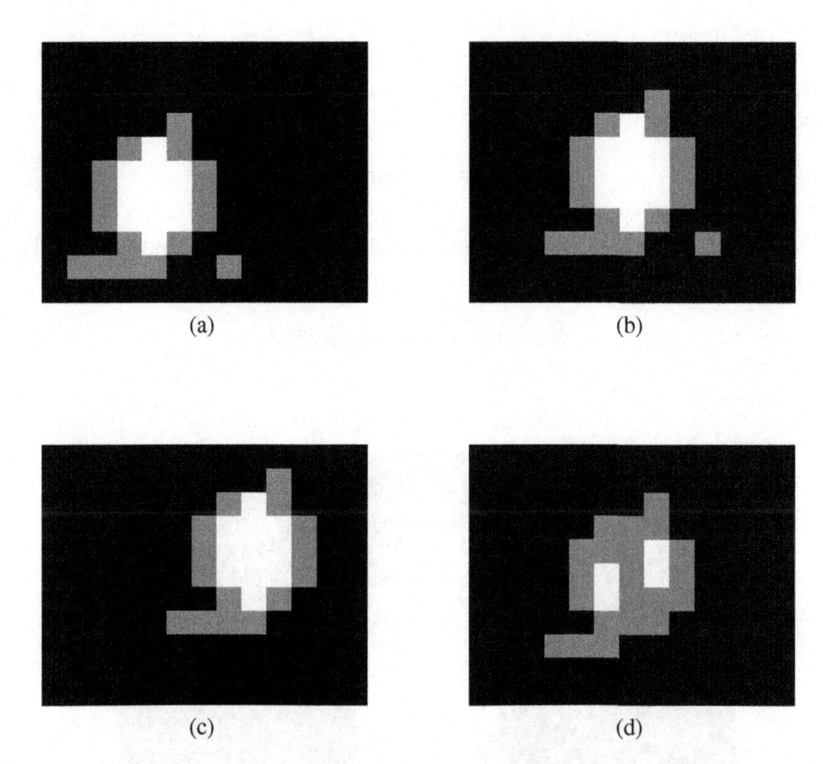

Fig. 7.2 (**a**) Shows a decision map D_1 in which the structures are contiguous. (**b**) and (**c**) Show decision maps D_2 and D_3. These are the same as D_1 but are slightly displaced up and down, left and right. (**d**) Shows the decision map $\widetilde{D}$ obtained by majority-vote fusion of D_1, D_2 and D_3. In this case, the fused map does not maintain the contiguous nature of the input maps.

7.7.1 Shape-Based Averaging

Shape-based averaging [8] was introduced specifically to address the above fragmentation problem. The basis of the algorithm is the *signed* distance transform which assigns to each pixel in the decision map its signed distance from the nearest "feature" pixel. If we regard any pixel with a label l as a feature pixel, then we may decompose a decision map $D_k(m,n)$ into L signed distance transforms $s_k(m,n|l), l \in \{1,2,\ldots,L\}$. Let $d_k(m,n|l)$ be the smallest Euclidean distance from (m,n) to a pixel with a label l and let $d_k(m,n|\widetilde{l})$ be the smallest Euclidean distance from (m,n) to a pixel with a label not equal to l, then the signed distance map $s_k(m,n|l)$ is defined as:

$$s_k(m,n|l) = d_k(m,n|l) - d_k(m,n|\widetilde{l}) \,. \tag{7.5}$$

According to (7.5) $s_k(m,n|l)$is negative if the pixel (m,n) lies inside the structure with label l, is positive if (m,n)lies outside the structure and is zero if, and only if, (m,n) lies on the perimeter of the structure (see Ex. 7.9). For each label $l, l \in \{1,2,\ldots,L\}$, we calculate a mean signed distance map, $\bar{s}(m,n|l)$, by averaging the $s_k(m,n|l)$ over all k:

$$\bar{s}(m,n|l) = \frac{1}{K}\sum_{k=1}^{K} s_k(m,n|l) .$$

The value of the fused decision map $\widetilde{D}(m,n)$ (Fig. 7.3) is then defined as the label l which has the minimum $\bar{s}(m,n|l)$ value:

$$\widetilde{D}(m,n) = \arg\min_{l} \bar{s}(m,n|l) .$$

Fig. 7.3 Shows the decision map $\widetilde{D}$ which is obtained by shape-based averaging the decision maps D_1, D_2 and D_3 which appear in Fig. 7.2(a)-(c). Observe how shape-based averaging helps to preserve the contiguous nature of the input images (cf. Fig. 7.2(d)).

Example 7.9. Signed Distance Transform. Consider the following one-dimensional image D wth three labels A, B and C:

$$D = (A\ A\ B\ C\ C\ C\ B\ C\ B\ A)^T .$$

The corresponding distance transforms $d(i|l=C)$, $d(i|\widetilde{l}=C)$ are:

$$d(i|l=C) = (3\ 2\ 1\ 0\ 0\ 0\ 1\ 0\ 1\ 2)^T ,$$
$$d(i|\widetilde{l}=C) = (0\ 0\ 0\ 1\ 2\ 1\ 0\ 1\ 0\ 0)^T ,$$

and the signed distance transform $s(i|l=C) = d(i|l=C) - d(i|\tilde{l}=C)$ is

$$s(i|l=C) = (3\ 2\ 1\ -1\ -2\ -1\ 1\ -1\ 1\ 2)^T\ .$$

7.7.2 *Similarity*

Decision maps are often fused together by measuring their similarity. This is often used in pattern recognition problems. Given two decision maps D_1 and D_2, we declare D_1 and D_2 to represent the same visual scene or object if the similarity measure $S(D_1,D_2)$ is greater than some threshold T.

Example 7.10. Face Recognition Using a Local Binary Pattern [1]. A direct method for performing face recognition is to compare a given test image B with a collection of training images $A_k, k \in \{1,2,\ldots,K\}$, which belong to K different individuals. In order to measure the similarity $S(B,A_k)$ we must ensure that the test image B and the training images $A_k, k \in \{1,2,\ldots,K\}$, are radiometrically calibrated. One way of doing this is to use the local binary pattern (LBP) operator (Sect. 3.4) to convert B into a decision map D_B and A_k into a decision map D_k.

7.7.3 *Label Permutation*

We now consider decision fusion when the $D_k, k \in \{1,2,\ldots,K\}$, are not semantically equivalent. In many cases we may assume that, to a good approximation, there is an unknown one-to-one correspondence between the labels in the different D_k. In other words we assume that each label p in D_k corresponds to a single label q in D_h and vice versa. In this case, we may simply solve the label correspondence problem by permuting the labels $p, p \in \{1,2,\ldots,L_k\}$, in D_k until the overall similarity

$$\sum_{k=1}^{K}\sum_{h=1}^{K} S(\pi_k(D_k), \pi_h(D_h))\ ,$$

is a maximum, where $\pi_k(D_k)$ denotes a permutation of the labels in D_k (see Ex. 5.5).

A convenient similarity measure for this purpose is the normalized mutual information NMI [9]:

$$NMI(D_k,D_h) = \sum_{p=1}^{L_k}\sum_{q=1}^{L_h} \widetilde{M}_{p,q} \log\frac{\widetilde{M}_{p,q}}{\widetilde{M}_p\widetilde{N}_q} \Big/ \sqrt{\sum_{p=1}^{L_k} \widetilde{M}_p \log(\widetilde{M}_p) \sum_{q=1}^{L_h} \widetilde{N}_q \log(\widetilde{N}_q)}$$

where $\widetilde{M}_p$ is the relative number of pixels in D_k with a label $p, p \in \{1,2,\ldots,L_k\}$, $\widetilde{N}_q$ is the relative number of pixels in D_h with a label $q, q \in \{1,2,\ldots,L_h\}$, and $M_{p,q}$ is the relative number of pixels which jointly have a label p in D_k and have a label q in D_q [1]. If $\widetilde{\pi}_k$ denotes the optimal permutation for D_k, then:

$$(\widetilde{\pi}_1,\widetilde{\pi}_2,\ldots,\widetilde{\pi}_K) = \arg \max_{\pi_1,\pi_2,\ldots,\pi_K} \sum_{k=1}^{K} \sum_{h=1}^{K} NMI(\pi_k(D_k),\pi_h(D_h)) \,. \tag{7.6}$$

Eq. (7.6) represents a difficult combinatorial optimization problem. However, greedy search techniques, including simulated annealing and genetic algorithm, may give an approximate solution in an acceptable time. Given the (approximate) optimal permutation $\widetilde{\pi}_k, k \in \{1,2,\ldots,K\}$, we may find $\widetilde{D}$ by applying the majority-vote rule to $\widetilde{\pi}_k(D_k)$:

$$\widetilde{D}(m,n) = l \qquad \text{if} \qquad \sum_{k=1}^{K} \delta\left(\widetilde{\pi}_k(D_k(m,n)),l\right) \geq \frac{K}{2} \,,$$

where

$$\delta(a,b) = \begin{cases} 1 \text{ if } a = b \,, \\ 0 \text{ otherwise} \,. \end{cases}$$

7.7.4 Co-associative Matrix

In this section we consider the fusion of K decision maps D_k when we do not, or cannot, solve the label correspondence problem: We suppose the D_k are of size $M \times N$ and are spatially aligned. We transform the D_k into a common representational format by converting them into co-associative matrices $A_k, k \in \{1,2,\ldots,K\}$:

$$A_k(i,j) = \begin{cases} 1 \text{ if } D_k(m_i,n_i) = D_k(m_j,n_j) \,, \\ 0 \text{ otherwise} \,. \end{cases}$$

Let $\widetilde{A}$ denote the result of fusing the A_k together. Then we define $\widetilde{D}$ as the decision map which corresponds to $\widetilde{A}$. This is illustrated in the next example.

Example 7.11. Mean Co-Association Matrix. The arithmetic mean is the simplest method for fusing co-association matrices:

$$\widetilde{A}(i,j) = \frac{1}{K} \sum_{k=1}^{K} A_k(i,j) \,.$$

[1] The relative number of pixels is a probability. Thus $\widetilde{M}_p = M_p/M$ and $\sum_p \widetilde{M}_p = 1$, where M_p is the number of pixels in D_k with a label p and M is the total number of pixels in D_k. Similarly, $\widetilde{N}_q = N_q/M$ and $\widetilde{M}_{p,q} = M_{p,q}/M$.

Given $\widetilde{A}$ we now search for a decision map $\widetilde{D}$ whose co-association matrix closely approximates $\widetilde{A}$. To do this we use spectral cluster algorithms. These algorithms use greedy search techniques and require an estimate of the number of clusters $\widetilde{L}$. A simple estimate of $\widetilde{L}$ [13] is:

$$\widetilde{L} = \min_{L}\left(\sum_{i=1}^{L} \widetilde{v}_i > \alpha \sum_{i=1}^{MN} \widetilde{v}_i\right) ,$$

where $\widetilde{v}_i$ is the ith largest eigenvalue of $\widetilde{A}$ and α is some fraction close to one. A reasonable value for α is $\alpha = 0.8$.

In Chapt. 20 we describe another method for fusing the $D_k, k \in \{1,2,\ldots,K\}$, which does not require solving the label correspondence problem and which does not use the co-association matrix.

7.8 Software

CLUSTERPACK. A matlab toolbox for spectral clustering. Authors: A. Strehl and J. Ghosh [9].

SPECTRAL CLUSTERING TOOLBOX. A matlab toolbox for spectral clustering. Authors: Deepak Verma and M. Meila. The toolbox may be used to cluster the mean co-associative matrix.

References

1. Ahonen, T., Hadid, A., Pietikainen, M.: Face description with local binary patterns: application to face recognition. IEEE Trans. Patt. Anal. Mach. Intell. 28, 2037–2041 (2006)
2. Bazi, Y., Bruzzone, L., Melgani, F.: Image thresholding based on the em algorithm and the generalized Gaussian distribution. Patt. Recogn. 40, 619–634 (2007)
3. Bertalmio, M., Caselles, V., Pardo, A.: Movie denoising by average of warped lines. IEEE Trans. Image Process. 16, 2333–2347 (2007)
4. Bruzzone, L., Prieto, D.F.: Automatic analysis of the difference image for unsupervised change detection. IEEE Trans. Geosci. Remote Sens. 38, 1171–1182 (2000)
5. Chung, K.-L., Lin, Y.-R., Huang, Y.-H.: Efficient shadow detection of color aerial images based on successive thresholding scheme. IEEE Trans. Geosci. Remote Sens. 47, 671–682 (2009)
6. Hesse, C.W., Holtackers, D., Heskes, T.: On the use of mixtures of Gaussians and mixtures of generalized exponentials for modelling and classification of biomedical signals. In: IEEE Benelux EMBS Symposium (2006)
7. Monwar, M.M., Gavrilova, M.L.: Multimodal biometric system using rank-level fusion approaches. IEEE Trans. Syst. Man Cybernetics 39B, 867–878 (2009)
8. Rohlfing, T., Maurer Jr., C.R.: Shape-based averaging. IEEE Trans. Image Process. 16, 153–161 (2007)

9. Strehl, A., Ghosh, J.: Cluster ensembles - a knowledge reuse framework for combining multiple partitions. J. Mach. Learn. Res. 3, 583–617 (2002)
10. Tsai, V.J.D.: A comparative study on shadow compensation of color aerial images in invariant color models. IEEE Trans. Geosci. Remote Sens. 44, 16671–16671 (2006)
11. Tu, T.-M., Su, S.-C., Shyu, H.-C., Huang, P.S.: A new look at IHS-like image fusion methods. Inf. Fusion 2, 177–186 (2001)
12. Wang, Z., Gao, C., Tian, J., Lia, J., Chen, X.: Multi-feature distance map based feature detection of small infra-red targets with small contrast in image sequences. In: Proc. SPIE, vol. 5985 (2005)
13. Wang, X., Yang, C., Zhou, J.: Spectral aggregation for clustering ensemble. In: Proc. Int. Conf. Patt. Recog. (2008)

Part II
Techniques

Chapter 8
Multi-resolution Analysis

Abstract. The subject of this chapter is multi-resolution analysis for images. We shall concentrate on the discrete wavelet transform (DWT) which provide a framework for the multi-resolution analysis of an input image by decomposing an input image into a sequence of wavelet planes and a residual image. We start by giving a brief review of multi-resolution analysis. We then move on to the DWT and its use in image fusion. To make our discussion more concrete we shall concentrate on two applications which rely on the DWT: (1) Fusion of an electro-optical image and an infra-red image. (2) Pan-sharpening in which we fuse a high spatial resolution panchromatic image with a low spatial resolution multi-spectral image.

8.1 Introduction

In multi-resolution analysis (MRA) we decompose an input image I into sequence of images $I_l, l \in \{1,2,\ldots,L\}$, each of which captures the information present in I at a given scale and orientation. Graphically, we may picture the I_l arranged in a pyramid (Fig. 8.1). At the bottom of the pyramid is the image I_0 which is identical to the input image I. At each successive level l the image I_l is recursively constructed by low-pass filtering and sub-sampling the image I_{l-1}. Given the approximation images I_l we create a sequence of detail images by interpolation of the approximation image I_l and subtraction of the outcome from its predecessor I_{l-1}.

By choosing appropriate low-pass filters we are able to select the change in resolution between the images I_l. In this chapter we shall assume dyadic MRA in which I_l has double the resolution of I_{l+1}.

Formally we define MRA as the mapping

$$(y_1, y_2, \ldots, y_L, I_L) = MRA(I) , \tag{8.1}$$

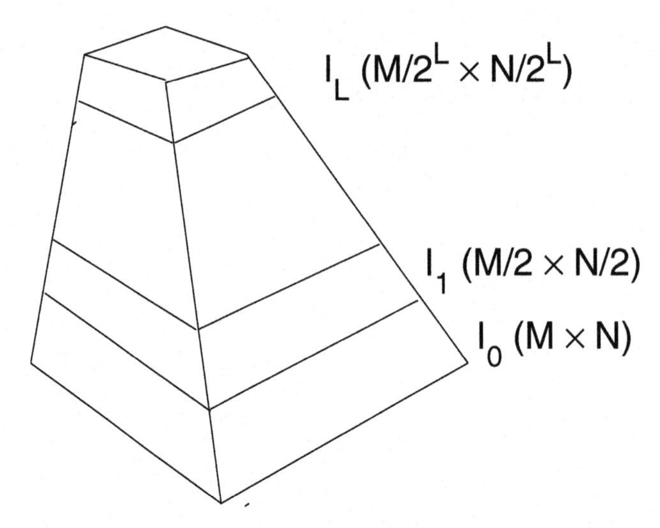

Fig. 8.1 Shows the pyramid representation of the multi-resolution analysis of an input image I. At the base of the pyramid is the $M \times N$ input image I. As we move up the pyramid, successive approximation images have an increasing coarser spatial resolution. At the lth level, the approximation image measures $M/2^l \times N/2^l$.

where y_l is the detail image at level l and I_L is the approximation at the coarsest resolution level L. The input image I may be recovered by applying the inverse MRA transformation:

$$I = MRA^{-1}(y_1, y_2, \ldots, y_L, I_L) . \tag{8.2}$$

8.2 Discrete Wavelet Transform

A special case of MRA is the discrete wavelet transform (DWT) decomposition in which the filters are specially designed so that successive layers of the pyramid only include details which are not already available at the preceeding levels. The DWT uses a cascade of special low-pass and high-pass filters and a sub-sampling operation.

We consider the decomposition of a one-dimensional signal x. The process of applying the DWT to x can be represented as a bank of filters, as in Fig. 8.2. At each level of decomposition, the signal x_l is split into a high-frequency component y_{l+1} and low-frequency component x_{l+1}. The low-frequency component x_{l+1} is then further decomposed until the desired resolution is reached.

Mathematically, the DWT decomposition of a one-dimensional signal x_l at level l is:

$$\begin{aligned} y_{l+1} &= \downarrow (x_l \otimes H) , \\ x_{l+1} &= \downarrow (x_l \otimes L) , \end{aligned}$$

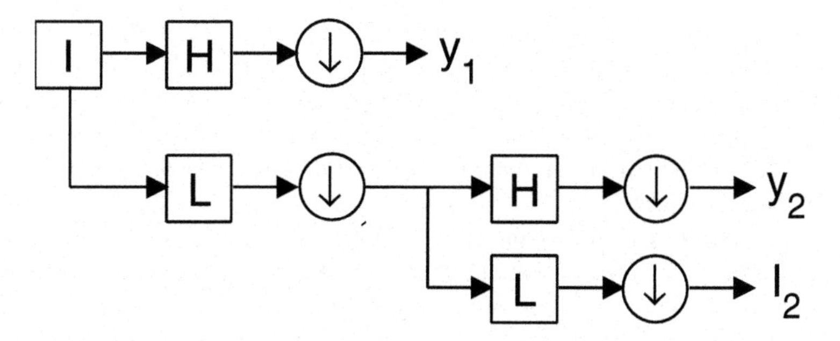

Fig. 8.2 Shows a two-level one-dimensional discrete wavelet transform. Input is a one-dimensional signal I. L and H denote, respectively, the low and high pass filters. $\downarrow$ denotes the operation of down-sampling by a factor of two. y_l and I_l are, respectively, the detail and approximate signals at the lth decomposition level.

where $\downarrow(x_l \otimes H)$ and $\downarrow(x_l \otimes L)$ denote, respectively, the convolution of x_l with the high-pass and low-pass filters H and L followed by sub-sampling the result by a factor of two.

The above procedure can be easily extended to a two-dimensional image as follows: Given a $M \times N$ input image I we generate two $(M/2) \times N$ images, I_L and I_H, by separately filtering and down-sampling the rows in I using a low-pass filter L and a high-pass filter H. We repeat the process by filtering and down-sampling the columns in I_L and I_H using the filters L and H. The output is four $(M/2) \times (N/2)$ images I_{LL}, I_{LH}, I_{HL} and I_{HH}, where I_{LL} is a low-frequency approximation of I, and I_{LH}, I_{HL} and I_{HH} are high-frequency detail images which represent horizontal (H), vertical (V) and diagonal (D) structures in I (Fig. 8.3).

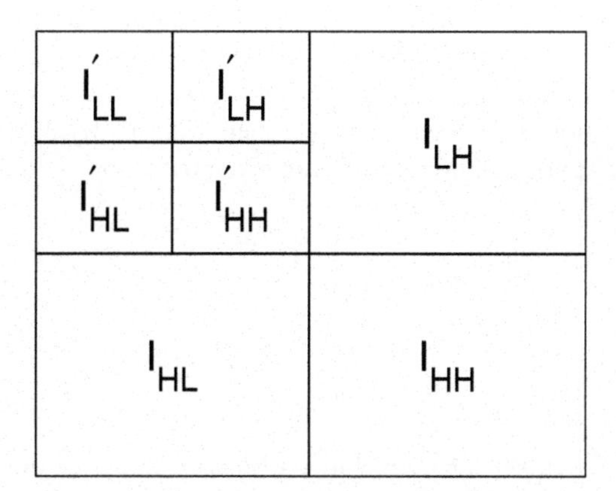

Fig. 8.3 Shows a $M \times N$ input image I decomposed into three $(M/2) \times (N/2)$ detail images I_{LH}, I_{HL} and I_{HH} and one $(M/2) \times (N/2)$ approximation image I_{LL}. The image I_{LL} is further decomposed into three $(M/4) \times (N/4)$ detail images I'_{LH}, I'_{HL} and I'_{HH} and one $(M/4) \times (N/4)$ approximation image I'_{LL}.

The corresponding DWT has relatively poor directional selectivity since it has only three characteristic directions: horizontal (0°), vertical (90°) and diagonal (45°). In some image fusion applications we require a transform with better directional selectivity. In this case, a convenient choice is the dual-tree complex wavelet transform (CWT) which has six characteristic directions: $15^o, 45^o, 75^o, 105^o, 135^o$ and 165° (see Ex. 13.1).

An important drawback to using the DWT in image fusion is the lack of shift invariance. This means that small shifts in the input images create unpredictable changes in the energy distribution of the detail image pixels. This in turn may lead to large distortions in the output. As a result, it is now common practice to use the dual-tree CWT (which is very nearly shift-invariant) or the undecimated DWT (which is exactly shift-invariant) [3, 6] for image fusion applications.

8.3 Undecimated Discrete Wavelet Transform (UDWT)

The undecimated discrete wavelet transform (UDWT) works by suppressing the down-sampling step in the DWT. Instead it up-samples the low-pass and high-pass filters by inserting zeros between the filter coefficients. Algorithms in which the filters are up-sampled are called *a trous* filters which means "with holes". Because there is no sub-sampling, the corresponding UDWT *a trous* detail images $y_l, l \in \{1, 2, \ldots, L\}$, and the approximation images I_l all have the same number of pixels as the input image I (see Fig. 8.4).

The following example illustrates a non-separable *a trous* decomposition, where for each level l, there is a single low-frequency approximation image I_l and a single high-frequency detail image y_l.

Example 8.1. Non-separable a Trous Decomposition. The non-separable *a trous* UDWT decomposition works as follows. Given an input image I, we construct a sequence of approximate images $I_1, I_2, \ldots, I_L$, by performing successive convolutions with a set of low-frequency masks $M_1, M_2 \ldots, M_L$. Mathematically, the approximation images are given by

$$
\begin{aligned}
I_1 &= M_1 \otimes I , \\
I_2 &= M_2 \otimes I_1 , \\
&\vdots \\
I_L &= M_L \otimes I_{L-1} .
\end{aligned}
$$

where $M_{l+1} \otimes I_l$ denotes the convolution of I_l with M_{l+1} and the mask M_{l+1} is derived from M_l by doubling its size and inserting zeros between the original values. Thus, if M_1 is the following 5×5 mask:

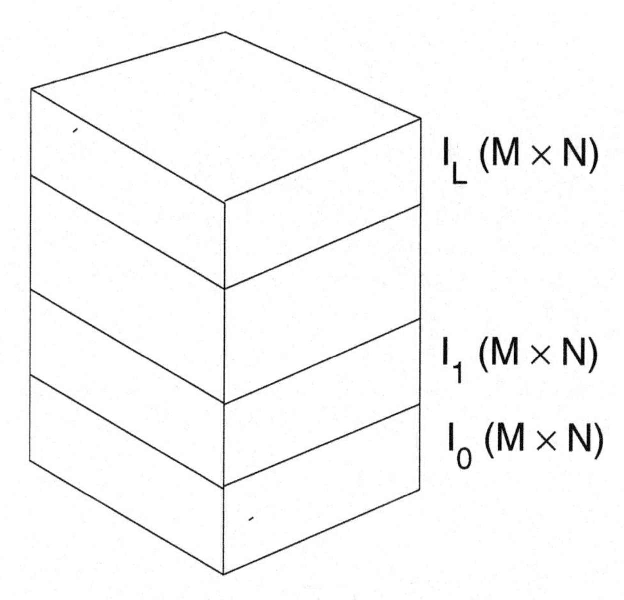

Fig. 8.4 Shows the parallelpiped representation of the UDWT decomposition of an input image I. At the base of the parallelpiped is the $M \times N$ input image I. As we move up the parallelpiped, successive approximation images have an increasing coarser spatial resolution. At all levels of the parallelpiped, the approximation images measure $M \times N$.

$$M_1 = \frac{1}{256}\begin{pmatrix} 1 & 4 & 6 & 4 & 1 \\ 4 & 16 & 24 & 16 & 4 \\ 6 & 24 & 36 & 24 & 6 \\ 4 & 16 & 24 & 16 & 4 \\ 1 & 4 & 6 & 4 & 1 \end{pmatrix}, \tag{8.3}$$

then M_2 is the following 9×9 mask

$$M_2 = \frac{1}{256}\begin{pmatrix} 1 & 0 & 4 & 0 & 6 & 0 & 4 & 0 & 1 \\ 0 & 0 & 0 & 0 & 0 & 0 & 0 & 0 & 0 \\ 4 & 0 & 16 & 0 & 24 & 0 & 16 & 0 & 4 \\ \vdots & & & & & & & & \\ 1 & 0 & 4 & 0 & 6 & 0 & 4 & 0 & 1 \end{pmatrix}.$$

The wavelet detail images $y_l, l \in \{1,2,\ldots,L\}$, are defined as the differences between two consecutive approximations I_{l-1} and I_l:

$$y_l = I_{l-1} - I_l ,$$

and the reconstruction formula for the input image I is

$$I = \sum_{l=1}^{L} y_l + I_r .$$

Fig. 8.5 shows the decomposition of an input image I into L detail images $y_l, l \in \{1,2,\ldots,L\}$, and L approximation images I_l.

The basic filter M_1 (8.3) and the corresponding derived masks $M_2, M_3, \ldots, M_L$ are widely used in the *a trous* algorithm. They are special masks which generate a non-directional (i. e. isotropic) dyadic decomposition of the input image I.

The *a trous* decomposition is widely used for feature extraction. The following two examples illustrate *a trous* edge detection and *a trous* spot detection.

Example 8.2. A Trous Wavelet Decomposition Applied to Image Edge Detection [15]. The *a trous* UDWT wavelet decomposition may be used directly to detect edges in an input image [15]. Given an input image I we perform a L wavelet decomposition. The wavelet images y_l contain high-frequency information and little low-frequency information. In this case, we may use the sum $\sum |y_l|$ as a simple but effective edge detector. In [15] the authors recommend using $L = 3$ detail images for this purpose (see Fig. 10.2).

Example 8.3. A Trous Wavelet Decomposition Applied to Image Spot Detection [11]. The *a trous* UDWT wavelet decomposition may be used directly to detect bright spots in a biological image I. Spots are small compared to I but are relatively large when analyzed locally. They are characterized by a small number of pixels with large gray-levels which are correlated across many wavelet levels. In this case, we may use the product $|\prod_{l=1}^{L} y_l|$ as a simple, but effective, spot detector.

8.4 Wavelet Fusion

The basic idea of wavelet fusion is as follows: Given a set of input images $I^{(k)}$ we decompose them into L detail images $y_1^{(k)}, y_2^{(k)}, \ldots, y_L^{(k)}$ and a single residual (approximation) image $I_L^{(k)}$:

$$(y_k^{(1)}, y_2^{(k)}, \ldots, y_L^{(k)}, I_L^{(k)}) = UDWT(I) .$$

Then at each decomposition level l we construct a composite detail image $\widetilde{y}_l$ by fusing the detail images $y_l^{(k)}, k \in \{1,2,\ldots,K\}$. At the coarsest resolution L we also construct a composite residual (approximation) image $\widetilde{I}_L$. In mathematical terms,

(a)

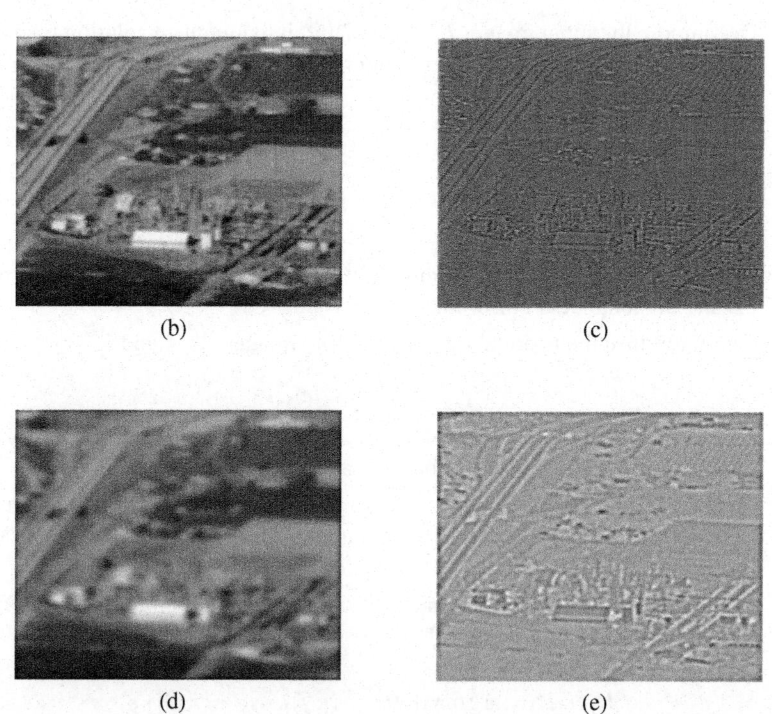

(b) (c)

(d) (e)

Fig. 8.5 **(a)** Shows an input image I. **(b)** and **(d)** Show the first and second approximation images I_1 and I_2. **(c)** and **(e)** Show the corresponding first and second detail images y_1 and y_2.

$$\widetilde{y}_l = f_l(y_l^{(1)}, y_l^{(2)}, \ldots, y_l^{(K)}) , \tag{8.4}$$

$$\widetilde{I}_L = g(I_L^{(1)}, I_L^{(2)}, \ldots, I_L^{(K)}) , \tag{8.5}$$

where f_l and g are appropriate fusion operators.

Given the composite detail images $\widetilde{y}_l, l \in \{1, 2, \ldots, L\}$, and the composite residual (approximation) image $\widetilde{I}_L$, we obtain the fused image I by applying the inverse UDWT transformation to $(\widetilde{y}_1, \widetilde{y}_2, \ldots, \widetilde{y}_L, \widetilde{I}_L)$:

$$\widetilde{I} = UDWT^{-1}(\widetilde{y}_1, \widetilde{y}_2, \ldots, \widetilde{y}_L, \widetilde{I}_L) .$$

This scheme [4, 8] was used to fuse together an electro-optical image and an infra-red image with the same spatial resolution. The main steps in the UDWT scheme of [4, 8] are given in the following example.

Example 8.4. Electro-Optical and Infra-Red Image Fusion [4, 8]. The main steps in the UDWT scheme for electro-optical and infra-red image fusion [4, 8] are:

1. Decompose the input images $I^{(EO)}$ and $I^{(IR)}$, into a multi-resolution format using a separable undecimated wavelet transform. We use only one decomposition level, i. e.

$$(y_1^{(EO)}, I_1^{(EO)}) = UDWT(I^{(EO)}) ,$$
$$(y_1^{(IR)}, I_1^{(IR)}) = UDWT(I^{(IR)}) ,$$

   where $y_1^{(k)} = (H_1^{(k)}, V_1^{(k)}, D_1^{(k)})$ and H, V and D denote the horizontal, vertical and diagonal detail information.
2. Optimally fuse the residual (approximation) images $I_1^{(EO)}$ and $I_1^{(IR)}$:

$$\widetilde{I}_1 = g(I_1^{(EO)}, I_1^{(IR)}) ,$$

   where we use the expectation-maximization (EM) algorithm for the fusion operator g: The residual images $I_1^{(EO)}$ and $I_1^{(IR)}$ are modeled using a Gaussian mixture model (GMM) whose parameters are optimized using the EM algorithm (see Sect. 8.5).
3. Fuse the detail images $y_1^{(EO)}$ and $y_1^{(IR)}$ using a simple maximum rule. For the pixel (i, j) the maximum rule is:

$$\widetilde{H}_1(i,j) = \arg\max(|H_1^{(EO)}(i,j)|, |H_1^{(IR)}(i,j)|) ,$$
$$\widetilde{V}_1(i,j) = \arg\max(|V_1^{(EO)}(i,j)|, |V_1^{(IR)}(i,j)|) ,$$
$$\widetilde{D}_1(i,j) = \arg\max(|D_1^{(EO)}(i,j)|, |D_1^{(IR)}(i,j)|) .$$

4. Perform the inverse undecimated DWT to obtain the fused image $\widetilde{I}$:

$$\widetilde{I} = UDWT^{-1}(\widetilde{H}_1, \widetilde{V}_1, \widetilde{D}_1, \widetilde{I}_1) .$$

8.5 Expectation-Maximization Algorithm

The EM algorithm is a general method for finding the maximum-likelihood estimates of an underlying distribution from a given data set in which the data is incomplete. In the context of fusing K images $I^{(k)}, k \in \{1,2,\ldots,K\}$, we suppose the $I^{(k)}$ are derived from an unknown image F. The fact that F is unknown implies that the data is incomplete. For example, in Ex. 8.4, $I^{(1)}$ is the residual (approximation) image $I_1^{(EO)}$ and $I^{(2)}$ is the residual (approximation) image $I_1^{(IR)}$.

A commonly used model for $I^{(k)}$ is:

$$I^{(k)}(i,j) = \alpha^{(k)}(i,j)F(i,j) + \beta^{(k)}(i,j) + \varepsilon^{(k)}(i,j) , \tag{8.6}$$

where $\alpha^{(k)}(i,j) \in \{-1,0,1\}$, is the sensor selectivity factor, $\beta^{(k)}(i,j)$ is the local bias of $I^{(k)}$ and $\varepsilon^{(k)}(i,j)$ is the random noise at (i,j). The sensor selectivity factor is used when the images do not have the same modality. In this case,

$$\alpha^{(k)}(i,j) = \begin{cases} 1 & \text{if } k\text{th sensor "sees" a given scene} , \\ 0 & \text{if } k\text{th sensor cannot "see" a given scene} , \\ -1 & \text{if } k\text{th sensor "sees" a given scene with reversed polarity} . \end{cases}$$

In the EM algorithm, we model the local noise $\varepsilon^{(k)}(i,j)$ as a mixture of M Gaussian probability density functions:

$$p(\varepsilon^{(k)}(i,j)) = \sum_{m=1}^{M} \frac{\lambda_m^{(k)}(i,j)}{\sqrt{2\pi}\sigma_m^{(k)}(i,j)} \exp -\frac{1}{2}\left(\frac{\varepsilon^{(k)}(i,j)}{\sigma_m^{(k)}(i,j)}\right)^2 . \tag{8.7}$$

To a first approximation, the sensor selectivity $\alpha^{(k)}(i,j)$, the bias $\beta^{(k)}(i,j)$ and the pdf parameters $\lambda_m^{(k)}(i,j)$ and $\sigma_m^{(k)}(i,j)$ are constant over a local neighborhood centered at (i,j). In this case, we may drop the pixel coordinates (i,j) and write instead $\alpha^{(k)}$, $\beta^{(k)}$, $\lambda_m^{(k)}$ and $\sigma_m^{(k)}$.

Assuming initial values for $\alpha^{(k)}$, $\beta^{(k)}$, $\lambda_m^{(k)}$ and $\sigma_m^{(k)}$ we calculate the best estimate for F. Then, using this estimate of F we calculate the maximum likelihood values for $\alpha^{(k)}$, $\beta^{(k)}$, $\lambda_m^{(k)}$ and $\sigma_m^{(k)}$. The iterative procedure is continued until values for $\alpha^{(k)}$, $\beta^{(k)}$, $\lambda_m^{(k)}$, $\sigma_m^{(k)}$ and F converge. For more details regarding the EM algorithm see Ex. 7.2 and Ex. 7.3 and [4, 8].

8.6 Multi-modal Wavelet Fusion

Burt and Kolczynsk [5] describe a general framework for multi-modal wavelet fusion (Fig. 8.6) which combines two different fusion models: selection and average. The overall fusion rule is determined by two measures: a local match measure $\rho_l(i,j)$ which determines which of the two models (selection or average) is to be employed and a local saliency measure $\sigma_l^{(k)}(i,j)$ which determines which detail image

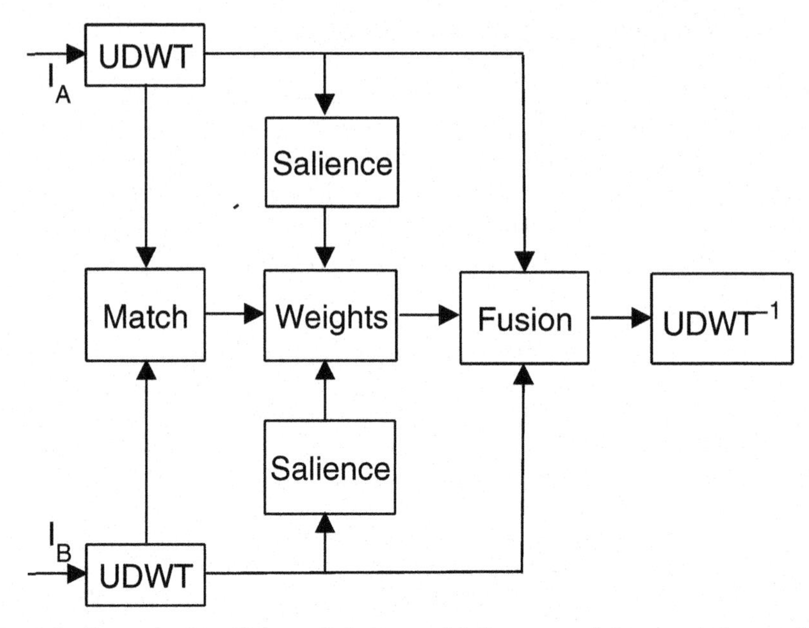

Fig. 8.6 Shows the Burt-Kolczynsk fusion model for a general framework for two input images I_1 and I_2

will be used (selection mode) or which detail image will be assigned a large weight (weighted average mode). Mathematically, the composite detail image at level l, $\widetilde{y}_l$, is given by:

$$\widetilde{y}_l(i,j) = \sum_k \Omega_k(i,j) y_l^{(k)}(i,j) ,$$

where the weights $\Omega_k(i,j)$ vary with the pixel location (i,j) and are a function of the local saliency measures $\sigma_l^{(k)}(i,j), k \in \{1,2,\ldots,K\}$, and the local matching coefficients $\rho_l^{(h,k)}(i,j), h,k \in \{1,2,\ldots,K\}, h \neq k$.

Let $\widetilde{y}_l^{(k)}(i,j|w), w \in \{1,2,\ldots,W\}$, denote W pixel values $y_l^{(k)}(i',j')$ which are in the local neighborhood of (i,j) (Fig. 8.7). Then

$$\sigma_l^{(k)}(i,j) = \sqrt{\frac{1}{W}\sum_{w=1}^{W} |\widetilde{y}_l^{(k)}(i,j|w) - \mu_l^{(k)}(i,j)|^2} ,$$

$$\rho_l^{(h,k)}(i,j) = \frac{1}{W}\sum_{w=1}^{W} (\widetilde{y}_l^{(h)}(i,j|w) - \mu_l^{(h)}(i,j))(\widetilde{y}_l^{(k)}(i,j|w) - \mu_l^{(k)}(i,j)) ,$$

where

$$\mu_l^{(k)}(i,j) = \frac{1}{W}\sum_{w=1}^{W} \widetilde{y}_l^{(k)}(i,j|w) .$$

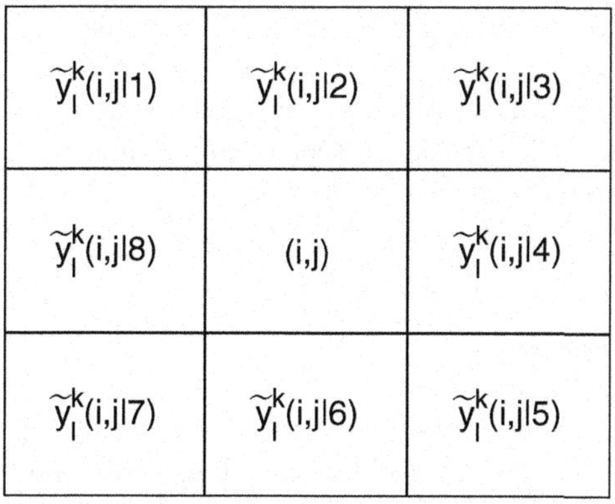

$\tilde{y}_l^k(i,j\|1)$	$\tilde{y}_l^k(i,j\|2)$	$\tilde{y}_l^k(i,j\|3)$
$\tilde{y}_l^k(i,j\|8)$	(i,j)	$\tilde{y}_l^k(i,j\|4)$
$\tilde{y}_l^k(i,j\|7)$	$\tilde{y}_l^k(i,j\|6)$	$\tilde{y}_l^k(i,j\|5)$

Fig. 8.7 Shows the pixel values $\tilde{y}_l^{(k)}(i,j|w)$ which are in a 3×3 neighborhood of (i,j), where $\tilde{y}_l^{(k)}(i,j|1) = y_l^{(k)}(i-1,j-1)$, $\tilde{y}_l^{(k)}(i,j|2) = y_l^{(k)}(i-1,j), \ldots, \tilde{y}_l^{(k)}(i,j|8) = y_l^{(k)}(i,j-1)$

For $K = 2$ inputs, the formulas for $\Omega_k(i,j)$ are:

$$\Omega_k(i,j) = \begin{cases} \Omega_{\max} & \text{if } \sigma_l^{(k)}(i,j) = \arg\max_h(\sigma_l^{(h)}(i,j)) \ , \\ \Omega_{\min} & \text{otherwise} \end{cases}$$

where

$$\Omega_{\min} = \begin{cases} \frac{1}{2}(1-(1-\rho_l)/(1-T)) & \text{if } \rho_l(i,j) > T \ , \\ 0 & \text{otherwise} \end{cases}$$

$$\Omega_{\max} = \begin{cases} \frac{1}{2}(1+(1-\rho_l)/(1-T)) & \text{if } \rho_l(i,j) > T \ , \\ 1 & \text{otherwise} \end{cases}$$

and $T = \frac{3}{4}$ is a threshold.

For the composite residual (approximation) image $\tilde{I}_L$ we may use the EM algorithm (Sect. 8.5). Alternatively, if the input images are of the same modality, we may use a simple arithmetic average operator:

$$\tilde{I}_L = \frac{1}{2}(I_L^{(1)} + I_L^{(2)}) \ .$$

Example 8.5. Generalized Gaussian Distribution [1]. Although very effective, the saliency and match formulas used by Burt and Kolczynsk assume the local distribution of the detail images are Gaussian. Experiments show that, in general, the $y_l^{(k)}(i,j)$ have heavier tails than a Gaussian distribution [2] and consequently, they are better described by a generalized Gaussian (GG) distribution:

$$p(y) = \frac{c_1(\alpha)}{\sigma} \exp - c_2(\alpha)(|y-\mu|/\sigma)^{\alpha} \ ,$$

where μ, σ and α are the mean, standard deviation and shape parameters of the GG distribution.

In this case, significantly higher quality fusion may be obtained if we use new saliency and match measures specifically designed for the generalized Gaussian distribution [2].

8.7 Pan-Sharpening

In this section we consider the application of wavelet fusion to pan-sharpening. Pan-sharpening attempts to create a high spatial resolution multi-spectral image by combining a high resolution panchromatic image and a low resolution multi-spectral image. To make our discussion more concrete we shall assume the multi-spectral image has only three spectral planes corresponding to the colors R, G and B and the ratio of the spatial resolution of the multi-spectral image to panchromatic image is $1:4$.

Pan-sharpening by means of the wavelet transform is based on the fact that the images $I_l, l \in \{0,1,\ldots,L\}$, are successive versions of the original image at increasing scales. Thus the first detail planes of the high-resolution panchromatic image contain spatial information which is not present in the multi-spectral image.

Example 8.6. Wavelet Additive Pan-sharpening [9, 10, 12]. In wavelet additive pan-sharpening we incorporate the high-resolution information directly into the intensity component of the multi-spectral image. Assuming a spatial resolution ratio of $1:4$, the steps in the algorithm are:

1. Spatially align the panchromatic and multi-spectral images and resample the multi-spectral image to make its pixel size equal to that of the panchromatic image.
2. Transform the multi-spectral image (RGB) into a (Iv_1v_2) image. This is a linear version of the intensity-hue-saturation transform (see Sect. 16.2.5):

$$\begin{pmatrix} I \\ v_1 \\ v_2 \end{pmatrix} = \begin{pmatrix} \frac{1}{3} & \frac{1}{3} & \frac{1}{3} \\ -\frac{\sqrt{2}}{6} & -\frac{\sqrt{2}}{6} & \frac{2\sqrt{2}}{6} \\ \frac{1}{\sqrt{2}} & -\frac{1}{\sqrt{2}} & 0 \end{pmatrix} \begin{pmatrix} R \\ G \\ B \end{pmatrix} .$$

3. Histogram-match the panchromatic image to the intensity image I. Let P denote the histogram-matched panchromatic image.
4. Decompose P into L detail images $y_l^{(P)}$ and a residual (approximation) image P_L:

$$(y_1^{(P)}, y_2^{(P)}, \ldots, y_L^{(P)}, P_L) = UDWT(P) .$$

5. Add the first two detail images of the panchromatic decomposition to the I component as follows:

$$\widetilde{I} = I + y_1^{(P)} + y_2^{(P)} . \tag{8.8}$$

6. Transform $(\widetilde{I} v_1 v_2)$ back to the RGB color space (see Sect. 16.2.5):

$$\begin{pmatrix} \widetilde{R} \\ \widetilde{G} \\ \widetilde{B} \end{pmatrix} = \begin{pmatrix} 1 & -\frac{1}{\sqrt{2}} & -\frac{1}{\sqrt{2}} \\ 1 & -\frac{1}{\sqrt{2}} & -\frac{1}{\sqrt{2}} \\ 1 & \sqrt{2} & 0 \end{pmatrix} \begin{pmatrix} \widetilde{I} \\ v_1 \\ v_2 \end{pmatrix} .$$

In substitute pan-sharpening we replace the first two detail images of the I component with the corresponding panchromatic detail images. Mathematically, we replace (8.8) with (8.9):

$$\widetilde{I} = I_2 + y_1^{(P)} + y_2^{(P)} , \tag{8.9}$$

where

$$(y_1^{(I)}, y_2^{(I)}, \ldots, y_L^{(I)}, I_L) = UDWT(I) .$$

8.8 Software

MATIFUS. A matlab toolbox for image fusion. Authors: P. M. de Zeeuw, G. Piella and H. J. A. M. Heijmans [14].

TOOLBOX-WAVELET. A matlab toolbox for wavelets including pyramid, *a trous* and curvelet decompositions. Author: Gabriel Peyre. Available from matlab central depository.

8.9 Further Reading

Refs. [7] and [13] review the use of wavelets in image fusion. A modern review and comparison of different wavelet fusion algorithms is [3].

References

1. Achim, A.M., Canagarajah, C.N., Bull, D.R.: Complex wavelet domain image fusion based on fractional lower order moments. In: Proc. 7th IEEE Int. Conf. Inform. Fusion (2005)
2. Achim, A.M., Loza, A., Bull, D.R., Canagarajah, C.N.: Statistical modelling for wavelet-domain image fusion. In: Stathaki, T. (ed.) Image Fusion: Algorithms and Applications. Academic Press, London (2008)
3. Amolins, K., Zhang, Y., Dare, P.: Wavelet based image fusion techniques - an introduction, review and comparison. ISPRS J. Photogramm. Remote Sens. 62, 249–263 (2007)
4. Blum, R.S., Yang, J.: Image fusion using the expectation-maximization algorithm and a Gaussian mixture model. In: Foresti, G.L., Regazzoni, C.S., Varshney, P.K. (eds.) Advanced video-based surveillance systems. Kluwer, Dordrecht (2003)

5. Burt, P.J., Kolczynski, R.J.: Enhanced image capture through fusion. In: Proc. 4th Int. Conf. Comp. Vis. (1993)
6. Chibani, Y., Houacine, A.: Redundant versus orthogonal wavelet decomposition for multisensor image fusion. Patt. Recogn. 36, 879–887 (2003)
7. Fenoy, G.P.: Adaptive wavelets and their applications to image fusion and compression. PhD thesis, University of Amsterdam (2003)
8. Liu, G., Jing, Z., Sun, S.: Image fusion based on an expectation maximization algorithm. Opt. Engn. 44, 077001-1–077001-11 (2005)
9. Nunez, J., Otazu, X., Fors, O., Prades, A., Pala, V., Arbiol, R.: Multi-resolution-based image fusion with additive wavelet decomposition. IEEE Trans. Geosci. Remote Sens. 37, 1204–1211 (1999)
10. Nunez, J., Otazu, X., Fors, O., Prades, A., Pala, V., Arbiol, R.: Image fusion with additive multiresolution wavelet decomposition. Applications to SPOT+Landsat images. J. Opt. Soc. Am. 16, 467–474 (1999)
11. Olivo-Marin, J.-C.: Extraction of spots in biological images using multiscale products. Patt. Recogn. 35, 1989–1996 (2002)
12. Otazu, X., Gonzlez-Audicane, M., Fors, O., Nunez, J.: Introduction of sensor spectral response into image fusion methods. Application to wavelet-based methods. IEEE Trans. Geosci. Remote Sens. 43, 2376–2385 (2005)
13. Pajares, G., de la Cruz, J.M.: A wavelet-based image fusion tutoral. Pattern Recogn. 37, 1855–1872 (2004)
14. Piella, G.: A general framework for multiresolution image fusion: from pixels to regions. Inf. Fusion 9, 259–280 (2003)
15. Zhang, X., Li, D.: A Trous Wavelet Decomposition Applied to Image Edge Detection. Geographic Inform. Sci. 7, 119–123 (2001)

Chapter 9
Image Sub-space Techniques

Abstract. The subject of this chapter is image sub-space techniques. These techniques are a special class of image transformations whose effect is project the input image into a lower dimensional space or sub-space. We shall concentrate on statistical sub-space methods which rely on a covariance matrix which is constructed from the input images. The techniques considered in this chapter include: principal component analysis (PCA), non-negative matrix factorization (NMF), canonical correlation analysis (CCA) and linear discriminant analysis (LDA).

9.1 Introduction

The basic idea of a subspace transformation is to project a high-dimensional input image into a lower dimensional space. Consider a $M \times N$ input image A. The image can be considered as a one-dimensional vector $\mathbf{a}$ of dimension MN, or equivalently, a point in a MN dimensional space. However, most natural images occupy only a small part of this space. In this case we should be able to find a low-dimensional sub-space in which a given natural image resides.

The main reasons for using a low dimensionality sub-space are:

Visualization. A reduction to a low-dimensional sub-space helps in understanding the intrinsic structure of the input data.

Generalization. A low-dimensional representation allows for better generalization.

Computational. In general manipulating low-dimensional data is both faster and requires less memory than the corresponding manipulations of high-dimensional data.

Model. The low-dimensional representation may be used as a model in its own right.

The following example describes the use of sub-space techniques in forming a low-dimensional space for face images.

Example 9.1. Face Space [14]. An ensemble of $M \times N$ face images maps to a collection of points in this MN-dimensional space. Images of faces, being similar in overall configuration, will not be randomly distributed in this space and thus can be described by a relatively low-dimensional "face space".

Mathematically a linear sub-space transformation works as follows. Given a column vector $\mathbf{a}$ of dimension MN, we may transform it into a column vector $\widetilde{\mathbf{a}}$ of reduced length L as follows:

$$\widetilde{\mathbf{a}} = W^T(\mathbf{a} - \phi) ,$$

where $W = (\mathbf{w}_1, \mathbf{w}_2, \ldots, \mathbf{w}_d)$ is a $MN \times d$ matrix which represents a set of d orthonormal basis functions $\mathbf{w}_i, i \in \{1,2,\ldots,d\}$, ϕ is a $MN \times 1$ column vector and $L \leq MN$.

Given $\widetilde{\mathbf{a}}$ we may recover an approximation of $\mathbf{a}$ by applying the inverse transformation as follows:

$$\hat{\mathbf{a}} = W\widetilde{\mathbf{a}} + \phi .$$

The following example describes the use of sub-space methods in face recognition.

Example 9.2. Sub-space Face Recognition. Sub-space methods are among the most successful approaches used in face recognition.. These methods project the input face into a low-dimensional sub-space where recognition is performed. Many different sub-space face recognitions algorithms have been proposed. They differ mostly in the kind of projection/decomposition method which is used and in the similarity matching criteria employed.

Depending on the required properties of the sub-space we obtain different sub-space representations. Among the different sub-space representations which are regularly used in image fusion are:

Principal component analysis (PCA). PCA is an unsupervised dimension reduction technique in which we seek an orthonormal basis function $W = (\mathbf{w}_1, \mathbf{w}_2, \ldots, \mathbf{w}_d)$ with $d << MN$, such that each individual image can be adequately represented as a linear combination of this basis. This requires that the error obtained when the input vector $\mathbf{a}$ is reconstructed from its low dimensional representation $\widetilde{\mathbf{a}}$ is minimal. We achieve this goal as follows. Given a training set of K input vectors $\mathbf{a}_k, k \in \{1,2,\ldots,K\}$, we seek directions which have the largest variances in the MN dimensional input space. The sub-space is reduced to a low dimension d by discarding those directions along which training vectors have a small variance.

Independent component analysis (ICA). ICA is an unsupervised dimension reduction technique which, contrary to PCA, not only finds uncorrelated components but it finds a linear transformation A such that the projections are as statistically independent as possible. ICA can be regarded as an extension of PCA, where the projections of the input data into the sub-space are not only uncorrelated but are also independent.

Canonical correlation analysis (CCA). CCA is an unsupervised dimension reduction technique which is used when there are two input vectors **a** and **b**. We seek a pair of directions such that the correlation between the projection of **a** along one direction and the projection of **b** along the second direction is a maximum. CCA is thus suitable for example for the estimation of orientation, where one set of observations consists of observed images, while the observations in the second set of object orientations from which the corresponding images were acquired.

Non-negative factorization (NMF). NMF is an unsupervised dimension reduction technique which is similar to PCA except the solution is constrained to have non-negative elements. Due to this non-negativity constraint, NMF tends to decompose the input images into parts leading to a part-based representation.

Linear discriminant analysis (LDA). LDA is a supervised dimension reduction technique in which we simultaneously maximize the distance between the projected class means and minimize the distances within classes.

Nearest neighbor discriminant analysis (NNDA). NNDA is a supervised dimension reduction technique in which we optimize the performance of a nearest neighbor classifier.

Kernel methods. All of the above dimension reduction techniques are linear. However, they may be made non-linear by employing the "kernel" trick: We use a non-linear mapping, or kernel, to map the input images into a high-dimensional feature space. We then perform one of the above linear dimension reduction techniques on the high-dimensional non-linear feature points. This procedure is mathematically equivalent to the applying a non-linear dimension reduction technique in the original input image space.

We shall start with PCA which is the probably the best known linear sub-space transformation.

9.2 Principal Component Analysis (PCA)

Principal component analysis (PCA) [9] seeks an orthonormal basis function $W = (\mathbf{w}_1, \mathbf{w}_2, \ldots, \mathbf{w}_d)$ with $d << MN$, such that each individual image can be adequately represented as a linear combination of this basis. Thus in face recognition, we seek the orthonormal vectors $\mathbf{w}_d$ which best account for the distribution of face images within the entire image space. These vectors define a sub-space of face images which we call "face space". Each of the vectors is of length MN, describes an $M \times N$ image and is a linear combination of the original face images. Because these vectors are the eigenvectors of the covariance matrix corresponding to the original face images, and because, after rearranging back into a rectangular image they are face-like in appearance, they are often referred to as "eigenfaces".

The orthonormal vectors $\mathbf{w}_i, i \in \{1,2,\ldots,d\}$, are found using a set of training images as follows: Let $A_k, k \in \{1,2,\ldots,K\}$, define a training set of face images each of size $M \times N$. We suppose the images A_k are all spatially aligned and radiometrically calibrated [1]. We rearrange the pixels in A_k into a column vector $\mathbf{a}_k$, where

$$\mathbf{a}_k = (a_k(1)a_k(2)\ldots a_k(MN))^T .$$

Then, the orthonormal vectors $\mathbf{w}_i, i \in \{1,2,\ldots\}$, are defined as the eigenvectors of the covariance matrix C, where

$$C = \frac{1}{K}\sum_{k=1}^{K}(\mathbf{a}_k - \phi)(\mathbf{a}_k - \phi)^T ,$$

$$\phi = \frac{1}{K}\sum_{k=1}^{K}\mathbf{a}_k ,$$

and

$$C\mathbf{w}_i = \lambda_i \mathbf{w}_i . \quad (9.1)$$

Since the column vectors $\mathbf{a}_k$ are of length MN, then C is a $MN \times MN$ matrix. In practice MN is so huge that eigenvector decomposition is computationally impossible. An alternative way to calculate the eigenvectors is as follows. Let

$$X = \left((\mathbf{a}_1 - \phi),(\mathbf{a}_2 - \phi),\ldots,(\mathbf{a}_K - \phi)\right) .$$

If $\mathbf{v}_i$ is the ith eigenvector of the matrix $X^T X$:

$$X^T X \mathbf{v}_i = \lambda_i \mathbf{v}_i ,$$

then pre-multiplying both sides of the equation with X, we obtain

$$XX^T X\mathbf{v}_i = CD\mathbf{v}_i = \lambda_i \mathbf{X}\mathbf{v}_i .$$

This means that if $\mathbf{v}_i$ is an eigenvector of the $K \times K$ matrix $X^T X$, then $\mathbf{w}_i = X\mathbf{v}_i$ is an eigenvector of the $MN \times MN$ covariance matrix $C = XX^T$. This is known as the Turk-Pentland algorithm.

The eigenvectors $\mathbf{w}_i$ are often called eigenfaces and are shown as $M \times N$ images in Fig. 9.1. Being the columns of a unitary matrix, the eigenfaces are orthogonal and efficiently span the space of a variation in the face images. Generally we select a small subset of $d << MN$ eigenfaces to define a low dimensional face-space that yields highest recognition performance on unseen examples of faces. For good recognition performance the required number of eigenfaces d is typically chosen to be of the order of 6 to 10.

[1] The spatial alignment and radiometric calibration of the input images is critical for face recognition. Often the spatial alignment algorithm relies on an eye localization and the radiometric calibration algorithm is based on histogram equalization. For an eye localization algorithm which is robust to variations in illumination, see [28].

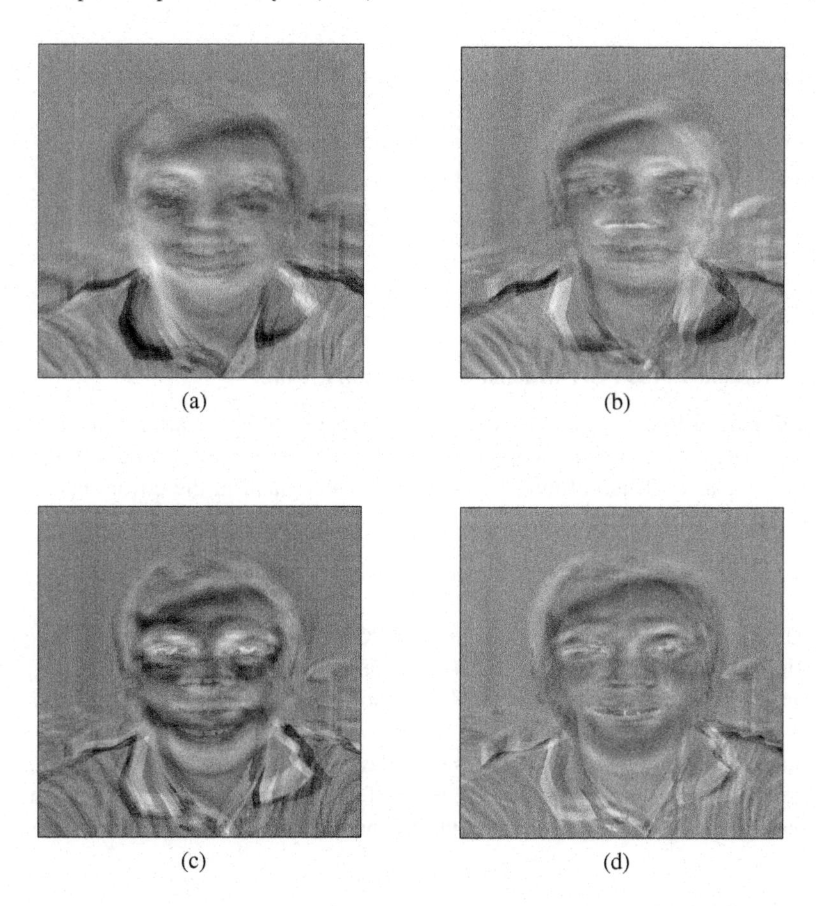

Fig. 9.1 **(a)-(d)** Shows the first four eigenimages generated from a training set of face images

Example 9.3. PCA face recognition [25]. In face recognition we wish to identify a given test image B. Off-line we select a set of K training images A_k which belong to several individuals. After spatial alignment and radiometric calibration we use (9.1) to generate a set of d eigenvectors $\mathbf{w}_i, i \in \{1,2,\ldots,d\}$. If we write B as a column vector $\mathbf{b}$, then we calculate feature vectors $\widetilde{\mathbf{b}}$ and $\widetilde{\mathbf{a}}_k, k \in \{1,2,\ldots,K\}$:

$$\widetilde{\mathbf{b}} = W^T(\mathbf{b} - \phi) ,$$
$$\widetilde{\mathbf{a}}_k = W^T(\mathbf{a}_k - \phi) .$$

We then use a nearest neighbor classifier to assign the test image B to a given individual. Let $D(\widetilde{\mathbf{b}}, \widetilde{\mathbf{a}}_k)$ be an appropriate distance measure between

$\widetilde{\mathbf{b}}$ and $\widetilde{\mathbf{a}}_k$. Then we classify the test image as belonging to the k^*th individual, where

$$k^* = \arg\min_k D(\widetilde{\mathbf{b}}, \widetilde{\mathbf{a}}_k) ,$$

9.2.1 PCA Variants

Some variants of the PCA which are in common use are:

Standardized PCA. In (9.1) we perform PCA using the covariance matrix C. In the context of remote sensing applications we often use the correlation matrix ρ in place of C. This is found by dividing each element in the covariance matrix C_{ij} by its standard deviation σ_{ij}:

$$\rho_{ij} = C_{ij}/\sigma_{ij} ,$$

This procedure is known as standardized PCA [23] and is often found [5] to yield higher signal-to-noise ratios in comparison with the conventional PCA.

Robust PCA. The conventional PCA is prone to the presence of outliers. This is because of how the covariance matrix C is calculated. For robust covariance matrix calculation see [8, 26] who have developed a robust PCA algorithm.

Class PCA. In class PCA [16, 17] we perform PCA separately on each class. If there are L classes, then the class PCA generates an ensemble of L sub-spaces.

9.2.2 Whitening

In many applications we equalize the dynamic range of the eigenvectors $\mathbf{u}_i$ by dividing the $\mathbf{u}_i$ through by $\sqrt{\lambda_i}$. Mathematically, we define the corresponding transformation as:

$$\Lambda^{-1/2} = \begin{pmatrix} \frac{1}{\sqrt{\lambda_1}} & 0 & \dots & 0 \\ 0 & \frac{1}{\sqrt{\lambda_2}} & \dots & 0 \\ 0 & 0 & \ddots & 0 \\ 0 & 0 & \dots & \frac{1}{\sqrt{\lambda_{MN}}} \end{pmatrix} .$$

The combination of PCA and the above normalization is referred to as whitening. Consider a $M \times N$ test image B. Let $\mathbf{b}$ denote the equivalent $MN \times 1$ column vector. Then the corresponding whitened vector is

$$\widetilde{\mathbf{b}}_W = \lambda^{-1/2}\widetilde{\mathbf{b}} = \Lambda^{-1/2}U^T(\mathbf{b}-\phi)\ ,$$

where $U = (u_1, u_2, \ldots, u_L)$. After whitening, the covariance matrix becomes an identity matrix which means $\widetilde{\mathbf{b}}_W$ is invariant to additional orthonormal transformations. After whitening we may rotate $\widetilde{\mathbf{b}}_W$ for maximum discriminant power [12].

9.2.3 Two-Dimensional PCA

By rearranging the pixels in A_k into a column vector $\mathbf{a}_k$ we have destroyed any row-to-row relationships that may exist between the pixel gray-levels. In an effort to overcome this drawback a two-dimensional PCA (2D-PCA) algorithm was invented [4, 6, 29]. This operates directly on the input images $A_k, k \in \{1, 2, \ldots, K\}$, without first converting them into column vectors $\mathbf{a}_k$. The algorithm works as follows: First we normalize each image by subtracting the average face Ψ:

$$X_k = A_k - \Psi\ .$$

Then we *define* an $N \times N$ image covariance matrix G as

$$G = \frac{1}{K}\sum_{k=1}^{K} X_k^T X_k\ .$$

Let $\mathbf{w}_i, i \in \{1, 2, \ldots, d\}$, define the set of N-dimensional eigenvectors of G corresponding to the d largest eigenvalues:

$$G\mathbf{w}_i = \lambda_i \mathbf{w}_i\ . \tag{9.2}$$

Then the corresponding low-dimensional, $M \times d$, representation of the $M \times N$ test image, $(B - \Psi)$, is

$$\widetilde{B} = (B - \psi)W\ ,$$

where

$$W = (\mathbf{w}_1, \mathbf{w}_2, \ldots, \mathbf{w}_d)\ .$$

The 2D-PCA transformation, as defined in (9.2), operates on the rows of the input image. Variants of the 2D-PCA algorithm include a "column" and a "diagonal" 2D-PCA transformation [35].

Example 9.4. DiaPCA [35]. Diagonal 2D-PCA is a variant of the 2D-PCA. Given a $M \times N$ input image B we may create a diagonal image D by shifting the rows of B to the right as shown in Fig. 9.2. The image D is then decomposed using the 2D-PCA algorithm.

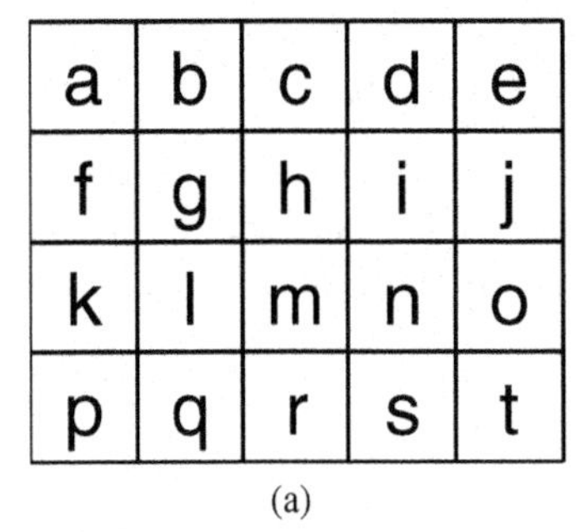

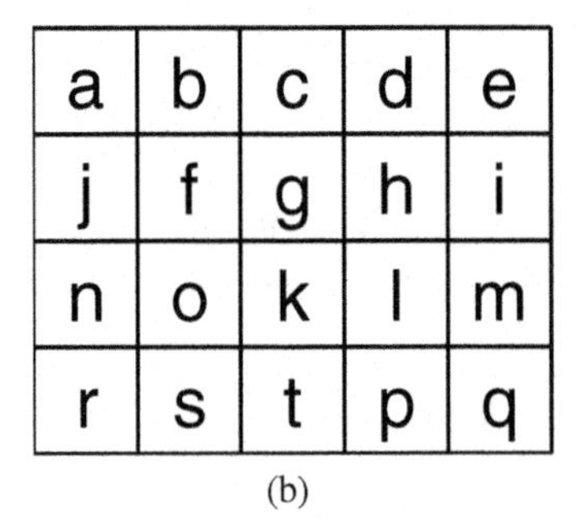

Fig. 9.2 **(a)** Shows a 4×5 image B. **(b)** Shows the corresponding diagonal image D formed by shifting rows of B to the right.

9.3 PCA Fusion

Most of the information in the input images $A_k, k \in \{1, 2, \ldots, K\}$, is present in the first eigenvector $\mathbf{w}_1$. The first eigenvector $\mathbf{w}_1$ (also known as the first principal component image) may therefore be regarded as the optimal fusion of the A_k into a single image. In general the fused image will be of less quality than any of the originals because we are only selecting the highest eigenvalue and therefore some of the structural patterns in the original images are lost. In order for PCA to be used effectively there needs to be a strong correlation between the original image data and the fused image data. An important PCA fusion technique is PCA pan-sharpening. Pan-sharpening is concerned with fusing a panchromatic image with a high spatial-resolution with a multi-spectral image with a low spatial resolution (see Chapt. 19).

In the traditional PCA pan-sharpening algorithm, we replace the first principal component image with the panchromatic image. The reason for this choice of component is that, by definition, the first principal component has the largest variance and therefore contains the most information. Before the first principal component is replaced by the panchromatic image, the panchromatic image is histogram matched (Sect. 6.2) to the first principal component image. The remaining principal components (eigenvectors) are considered to have band-specific information and are left unaltered. The inverse PCA is performed on the modified PAN image and the principal components to obtain a high-resolution pan-sharpened image.

An improved PCA pan-sharpened image may be obtained by optimally choosing which principal component to replace. In [24] we replace the principal component which has the highest correlation with the panchromatic image. The complete pan-sharpened procedure is as follows:

Example 9.5. Modified PCA Pan-Sharpening [24].

1. Perform PCA on the multi-spectral image.
2. Calculate the cross-correlation between the principal components and the panchromatic image.

3. Select the principal component having the highest absolute value of the correlation coefficient.
4. Inverse the panchromatic image before performing histogram matching if the cross-correlation coefficient is negative.
5. Perform histogram matching of the selected principal component and the panchromatic image
6. Use the histogram matched panchromatic image in the PCA based method for substitution or injection of the high spatial detail

9.4 Non-negative Matrix Factorization (NMF)

Mathematically, non-negative matrix factorization (NMF) [2, 7] is similar to PCA. Given the training images $A_k, k \in \{1,2,\ldots,K\}$, we write them as $MN \times 1$ column vectors $\mathbf{a}_k$. Let

$$X = (\mathbf{a}_1, \mathbf{a}_2, \ldots, \mathbf{a}_K) ,$$

Then, in NMF we approximate X by the multiplication of two matrices W and H:

$$X \approx WH , \tag{9.3}$$

where W is a non-negative $MN \times r$ mixing matrix and H is a non-negative $r \times K$ encoding matrix. The parameter r determines the accuracy of the approximation in (9.3).

Example 9.6. NMF Face Recognition [7]. Given a training set of K spatially aligned and radiometrically calibrated images $A_k, k \in \{1,2,\ldots,K\}$, we calculate a NMF mixing matrix W and a NMF encoding matrix H using (9.3). The NMF representation of each training image A_k is

$$\widetilde{\mathbf{a}}_k = W^{-1}\mathbf{a}_k .$$

Given a test image B, we write it as a column vector $\mathbf{b}$. Then its NMF representation is

$$\widetilde{\mathbf{b}} = W^{-1}\mathbf{b} .$$

We classify B as belonging to the k^*th individual, where

$$k^* = \arg\min_k \left(D(\widetilde{\mathbf{b}}, \widetilde{\mathbf{a}}_k)\right) ,$$

and $D(\widetilde{\mathbf{b}}, \widetilde{\mathbf{a}}_k)$ is an appropriate distance similarity measure.

Traditionally the matrices W and H are found by minimizing the square distance $||X - WH||^2$. Recently [22] have described a NMF algorithm in which we find W

and H by simultaneously minimizing the earth movers's distance (EMD) (see Sect. 14.2.3) between X and WH:

$$(W_{EMD}, H_{EMD}) = \arg\min_{(W,H)} \big(EMD(X, WH)\big) .$$

For pattern recognition applications, the new NMF has proved to be much more robust against errors in spatial alignment and radiometric calibration.

9.5 Linear Discriminant Analysis (LDA)

Linear discriminant analysis (LDA) finds a set of projection vectors which best discriminant between the different classes which are represented in the input data. Given a set of K input images $A_k, k \in \{1,2,\ldots,K\}$, or column vectors $\mathbf{a}_k$, we suppose each $\mathbf{a}_k$ belongs to a class $l, l \in \{1,2,\ldots,L\}$, where

$$y_k = l \qquad \text{if } \mathbf{a}_k \text{ belongs to class } l .$$

The LDA computes a linear transformation $W = (\mathbf{w}_1, \mathbf{w}_2, \ldots, \mathbf{w}_{L-1})$ which maps the $\mathbf{a}_k$ to an $(L-1)$-dimensional space:

$$\widetilde{\mathbf{a}}_k = W\mathbf{a}_k ,$$

in which the classes are maximally separated. Mathematically, W is defined as follows: Let S_B and S_W denote, respectively, the between-class and the within-class scatter matrices:

$$\begin{aligned} S_B &= \sum_{l=1}^{L} p_l(\mu_l - \mu)(\mu_l - \mu)^T = \sum_{l=1}^{L-1}\sum_{k=l+1}^{L} p_l p_k(\mu_l - \mu_k)(\mu_l - \mu_k)^T , \\ S_W &= \sum_{l=1}^{L} p_l S_l , \\ \mu &= \sum_{l=1}^{L} p_l \mu_l , \end{aligned}$$

where S_l is the scatter matrix for column vectors which belong to the lth class and μ_l and p_l are, respectively, the mean vector and the *a priori* probability of the samples in the lth class [2]. Then LDA minimizes the ratio of the determinant of the transformed scatter matrices $\widetilde{S}_B$ to $\widetilde{S}_W$:

$$W = \arg\max_W \frac{|\widetilde{S}_B|}{|\widetilde{S}_W|} = \arg\max_W \frac{|W^T S_B W|}{|W^T S_W W|} . \tag{9.4}$$

[2] Maximum likelihood estimate of p_l is the number of training saples which belong to class l divided by the total number of training samples.

Eq. (9.4) is known as the Fisher criterion. If S_W is a full rank matrix, then the solution of (9.4) is $W = (\mathbf{w}_1, \mathbf{w}_2, \ldots, \mathbf{w}_{L-1})$, where $\mathbf{w}_l$ is the eigenvector of $S_W^{-1} S_B$ with the lth largest eigenvalue:

$$(S_W^{-1} S_B)\mathbf{w}_l = \lambda_l \mathbf{w}_l \ . \tag{9.5}$$

However, if the number of input images, K, is small compared to their dimensionality [3], S_W becomes singular and we cannot use (9.5) to find the transformation W.

9.5.1 *Fisherface*

A common approach to finding the transformation W when S_W is singular is to simply project the input images A_k into a sufficiently low-dimensional space using PCA. In the reduced PCA space S_W is no longer singular and we may calculate the LDA transformation using (9.5). The technique is known as Fisherface [1]. Mathematically, the transformation is given by:

$$W_{fisherface} = W_{PCA} W_{LDA} \ .$$

Alternative methods [34] which avoid the singularity in S_W include:

Total Scatter Matrix. We replace the within-class scatter matrix S_W in (9.4) by the total scatter matrix $S_T = S_W + S_B$. In this case the transformation matrix $W_T = (\mathbf{w}_1, \mathbf{w}_2, \ldots, \mathbf{w}_{L-1})$, where

$$(S_T^{-1} S_B)\mathbf{w}_l = (S_W + S_B)^{-1} S_B \mathbf{w}_l = \lambda_l \mathbf{w}_l \ .$$

When S_W is non-singular the transformation matrix W_T is identical to the transformation W obtained using (9.4). When S_W is singular the modified criterion reaches the maximum value for any transformation W in the null space of S_W. The transformation W cannot guarantee the class separability, since $|W^T S_B W|$ is maximum for all W.

Null LDA. We work only in the null space of S_W. In this space, the transformation matrix W_{NLDA} is defined as the transformation which maximizes $|W^T S_B W|$ [3].

Direct LDA. We work only in the space which lies outside the null space of S_B. In this space the transformation W_{DLDA} is defined as the transformation which optimizes the Fisher criterion (9.4) [18, 31].

All of these methods, including Fisherface are, however, sub-optimal because they discard important discriminative information. The loss of discriminative information may, however, be reduced by using an ensemble of LDA transformations (see Chapt. 10).

[3] In all practical cases the number of input images, K, is much smaller than the size of the input images, $M \times N$.

Example 9.7. Random Sampling LDA [27]. We apply PCA to a training set of K column vectors $\mathbf{a}_k, k \in \{1,2,\ldots,K\}$. The eigenvectors with zero eigenvalues are removed and we retain the remaining $(K-1)$ eigenvectors $U = (\mathbf{u}_1,\mathbf{u}_2,\ldots,\mathbf{u}_{K-1})$. We use the following technique (see Sect. 10.3.1) to generate R random sub-spaces. Each sub-space is spanned by $\alpha+\beta$ dimensions. The first α dimensions are fixed as the α largest eigenvectors in U. The remaining β dimensions are randomly selected from the other $(K-1-\alpha)$ eigenvectors in U. For each random sub-space we construct an LDA classifier. Thus altogether we have an ensemble of K LDA classifiers.

By combining the classifiers (see Chapt. 10) we able to construct a powerful classifier that covers the entire feature space without losing discriminant information.

9.5.2 Median LDA

The LDA is based on non-robust estimates of population parameters. This means the LDA is sensitive to outliers. If outliers are thought to be present, the population parameters should be calculated using robust techniques. The following is a simple robust LDA algorithm.

In the median LDA [30] we replace the mean vectors μ_k by the corresponding median vectors $\mathbf{m}_k$. The corresponding between-class and within-class scatter matrices are:

$$S_B = \sum_{k=1}^{K} p_k(\mathbf{m}_k - \mathbf{m})(\mathbf{m}_k - \mathbf{m})^T \; ,$$

$$S_W = \sum_{l=1}^{L} p_l S_l \; ,$$

$$\mathbf{m} = \sum_{l=1}^{L} \mathbf{m}_l \; ,$$

where S_l is a robust scatter matrix calculated for column vectors $\mathbf{a}_k$ which belong to the lth class. In calculating a robust estimate for S_l we use the median vector $\mathbf{m}_l$ in place of the arithmetic mean vector μ_l. In addition [30] recommends weighing $\mathbf{a}_k$ according to their distance from $\mathbf{m}_l$.

9.5.3 Re-weighting LDA

The LDA transformation matrix W is invariant to any scale variation of the vectors $\mathbf{w}_l$. If $\mathbf{w}_l$ is a solution of (9.5), then $\alpha_l \mathbf{w}_l$ is also a solution. Although all eigenvectors $\alpha_l \mathbf{w}_l$ are optimum with regard to the Fisher criterion, they may not be optimal for classification purposes.

Given the transformation $W = (\mathbf{w}_1, \mathbf{w}_2, \ldots, \mathbf{w}_{L-1})^T$, we may learn a new transformation matrix

$$W_{reweight} = (\alpha_1 \mathbf{w}_1, \alpha_2 \mathbf{w}_2, \ldots, \alpha_{L-1} \mathbf{w}_{L-1}) ,$$

where the weights $\alpha_l, l \in \{1, 2, \ldots, L-1\}$, are learnt for optimal classification [13].

9.5.4 Two-Dimensional LDA

Following the development of the 2D-PCA algorithm, a two-dimensional version of the LDA algorithm was developed. The 2D-LDA algorithm [4, 11] operates directly on the input images $A_k, k \in \{1, 2, \ldots, K\}$, without first converting them into column vectors $\mathbf{a}_k$. The algorithm works by formally defining between-class and within-class image scatter matrices:

$$G_B = \sum_{l=1}^{L} p_l (\bar{A}_k - \bar{A})(\bar{A}_k - \bar{A})^T \quad \text{and} \quad G_W = \sum_{l=1}^{L} p_l G_l ,$$

where G_l is the image scatter matrix for the images A_k which belong to the lth class, and

$$\bar{A}(m,n) = \frac{1}{K} \sum_{k=1}^{K} A_k(m,n) .$$

Let $\mathbf{w}_i, i \in \{1, 2, \ldots, d\}$, define the set of M-dimensional eigenvectors of $G_W^{-1} G_B$ corresponding to the d largest eigenvalues:

$$G_W^{-1} G_B \mathbf{w}_i = \lambda_i \mathbf{w}_i . \tag{9.6}$$

Then the corresponding $d \times N$ projection of a $M \times N$ test image B is

$$\widetilde{B} = W^T B .$$

where $W = (\mathbf{w}_1, \mathbf{w}_2, \ldots, \mathbf{w}_d)$.

The 2D-LDA algorithm as defined in (9.6) operates on the columns of the input image. Variants of the 2D-LDA algorithm include a "row" 2D-LDA transformation [19].

The following example illustrates how the 2D-PCA and 2D-LDA techniques may be combined in a joint $(2D)^2$PCALDA classifier.

Example 9.8. $(2D)^2$PCALDA Classifier [19]. A novel method for image feature extraction is to simultaneously apply the 2D-PCA transformation W_{2DPCA} (9.2) and the 2D-LDA transformation W_{2DLDA} (9.6) on a $M \times N$ test image B. The result is a $d \times d$ feature matrix F:

$$F = W_{2DLDA}^T (B - \Psi) W_{2DPCA} . \tag{9.7}$$

The matrix F contains both the discriminant information of B extracted by the column-based 2D-LDA algorithm and the descriptive information of B extracted by the row-based 2D-PCA algorithm.

The feature matrix F may be used in a nearest neighbor classifier as follows. During training, each training image $A_k, k \in \{1,2,\ldots,K\}$, is projected onto both W_{2DPCA} and W_{2DLDA} simultaneously to obtain the respective feature matrix F_k. Let B be a given test image. We use (9.7) to find the corresponding feature matrix F. We then classify B using a nearest neighbor classifier:

$$y = y_k \qquad \text{if} \qquad d(F,F_k) = \min_h (d(F,F_h)) ,$$

where y_k is the class label of A_k and $D(F,F_h)$ is an appropriate distance measure (see Chapt. 14) defined between F and F_h.

9.6 Nearest Neighbor Discriminant Analysis (NNDA)

The LDA guarantees the optimum discriminative transformation when the class distributions are unimodal and are separated by the scatter of the class matrices. However, if the class distributions are multi-modal and share the same mean, the LDA fails to find the optimal discriminative transformation. The nearest neighbor discriminant analysis (NNDA) technique [20] was developed to help overcome these drawbacks.

In NNDA we find a linear transformation matrix W which optimizes the performance of a nearest neighbor classifier. Given a set of K $M \times N$ images $A_k, k \in \{1,2,\ldots,K\}$, or column vectors $\mathbf{a}_k$, we suppose each $\mathbf{a}_k$ belongs to a class $l, l \in \{1,2,\ldots,L\}$, where

$$\mathbf{a}_k = (a_k(1), a_k(2), \ldots, a_k(MN))^T ,$$

and

$$y_k = l \qquad \text{if } \mathbf{a}_k \text{ belongs to class } l .$$

Then NNDA finds a set of projection vectors, $W = (\mathbf{w}_1, \mathbf{w}_2, \ldots, \mathbf{w}_{MN})$, which maximizes the difference between the transformed scatter matrices, $\widetilde{C}_B$ and $\widetilde{C}_W$:

$$W = \arg\max_W |\widetilde{C}_B - \widetilde{C}_W| = \arg\max_W |W^T(C_B - C_W)W| , \qquad (9.8)$$

where

$$C_B = \sum_{k=1}^{K} (\mathbf{a}_k - \mathbf{x}_k)(\mathbf{a}_k - \mathbf{x}_k)^T \qquad \text{and} \qquad C_W = \sum_{k=1}^{K} (\mathbf{a}_k - \mathbf{y}_k)(\mathbf{a}_k - \mathbf{y}_k)^T ,$$

and $\mathbf{x}_k$ is defined as the column vector which is nearest to $\mathbf{a}_k$ and belongs to the same class as $\mathbf{a}_k$ and $\mathbf{y}_k$ is defined as the column vector which is nearest to $\mathbf{a}_k$ and which does not belong to the same class as $\mathbf{a}_k$.

9.6.1 K-Nearest Neighbor Discriminant Analysis

The NNDA transformation matrix W as defined in (9.8) optimizes the performance of a nearest neighbor classifier. We may generalize (9.8) to the case of L nearest neighbors by simply redefining $\mathbf{x}_k$ and $\mathbf{y}_k$ as follows: $\mathbf{x}_k$ is now defined as the column vector which is the $(L/2)$th nearest neighbor of $\mathbf{a}_k$ and belongs to the same class as $\mathbf{a}_k$ and $\mathbf{y}_k$ is now defined as the column vector which is the $(L/2+1)$th nearest neighbor of $\mathbf{a}_k$ and does not belong to the same class as $\mathbf{a}_k$.

9.6.2 Two-Dimensional NNDA

Ref. [21] describes a two-dimensional version of the NNDA algorithm (2D-NNDA). This works directly on the input images $A_k, k \in \{1,2,\ldots,K\}$, without first converting them into column vectors $\mathbf{a}_k$. The development of the 2D-NNDA algorithm parallels that of the NNDA algorithm. Thus the 2D-NNDA algorithm uses images X_K and Y_k instead of column vectors $\mathbf{x}_k$ and $\mathbf{y}_k$. The images X_k and Y_k are defined as follows: X_k is the input image which is closest to A_k (using an appropriate distance measure) and belongs to the same class as A_k and Y_k is the input image which is closest to A_k but does not belong to the same class as A_k.

9.7 Canonical Correlation Analysis (CCA)

Canonical correlation analysis (CCA) is a powerful method for comparing two multi-spectral images A and B when the images do not have the same number of spectral bands. In what follows we shall assume A has K spectral bands: $A = (A_1, A_2, \ldots, A_K)$ and B has L spectral bands: $B = (B_1, B_2, \ldots, B_L)$.

Canonical correlation analysis works as follows. We transform the multi-spectral images A and B into two scalar images $\widetilde{A}$ and $\widetilde{B}$ by making linear combinations of their spectral bands. Mathematically,

$$\widetilde{A}(x,y) = \sum_{k=1}^{K} \alpha_k A_k(x,y) \quad \text{and} \quad \widetilde{B}(x,y) = \sum_{l=1}^{L} \beta_l B_l(x,y) \; .$$

The vectors $\alpha = (\alpha_1, \alpha_2, \ldots, \alpha_K)$ and $\beta = (\beta_1, \beta_2, \ldots, \beta_L)$ are chosen so that the correlation $\rho(\widetilde{A}, \widetilde{B})$ between $\widetilde{A}$ and $\widetilde{B}$ is maximized, where

$$\rho(\widetilde{A}, \widetilde{B}) = \frac{\alpha^T S_{AB} \beta}{\sqrt{\alpha^T S_A \alpha \beta^T S_B \beta}} \; .$$

The idea is that we find the maximum difference between the multi-spectral images A and B after removing correlations between them as much as possible. Mathematically, α and β are found by solving the following coupled generalized eigenvector equations:

$$S_{AB}S^{-1}\beta S_{BA} = \rho^2 S_A \alpha \, ,$$
$$S_{BA}S^{-1}\alpha S_{AB} = \rho^2 S_B \beta \, ,$$

where S_A and S_B are the covariance matrices of the two images and $\Sigma_{AB} = \Sigma_{BA}^T$ is the inter-image covariance matrix. Solutions of of the eigenvector equations generate new spectral images $\mathbf{u} = (u_1, u_2, \ldots, u_K)^T$ and $\mathbf{v} = (v_1, v_2, \ldots, v_L)^T$ where the components u_k, v_k are known as canonical variates.

9.8 Software

CCA. A matlab routine for canonical correlation analysis. Author: Magnus Borga.

EIGENFACES FOR EXPRESSION DETECTION. A matlab toolbox for classifying different human expressions using PCA. Available from Matalb central directory. Author: Iftekhar Tanveer.

LIBRA. A matlab toolbox for classical and robust statistics [26]. The toolbox contains classical and robust implementations of several sub-space techniques. Authors: Sabine Verboven and Mia Hubert.

ROBCOEFF. A matlab toolbox for robust estimation of PCA, LDA and CCA. Authors: Daniel Skocaj, Ales Leonardis and Sanja Fidler.

STATISTICAL LEARNING TOOLBOX. A matlab statistical learning toolbox. the toolbox includes algorithms for several sub-space techniques. Author: Dahua Lin.

TOOLBOX-DIMREDUC. A matlab toolbox for dimensionality reduction methods. The toolbox includes algorithms for many sub-space techniques. Author: Gabriel Peyre.

9.9 Further Reading

Recently, 2D versions of the PCA, NMF, LDA, NNDA and CCA transformations have been developed [7, 10, 15, 32]. In general, however, the performance of the 2D techniques is not significantly different from that of the traditional 1D techniques [11]. A major problem with sub-space methods generally is their sensitivity to alignment. For a review of this see [33].

References

1. Belhumer, P.N., Hespanha, J.P., Kriegman, D.J.: Eigenfaces vs. Fisherfaces: recognition using class specific linear projection. IEEE Trans. Patt. Anal. Mach. Intell. 19(9), 711–720 (1997)

2. Buciu, I.: Non-negative matrix factorization. A new tool for feature extraction: Theory and Applications. Int. J. Comput. Comm. Control 3, 67–74 (2008)
3. Chen, L., Liao, H., Ko, M., Lin, J., Yu, G.: A new LDA-based face recognition system which can solve the small sample size problem. Patt. Recogn. 33, 1713–1726 (2000)
4. Chen, S., Zhu, Y., Zhang, D., Yang, J.-Y.: Feature extraction approaches based on matrix pattern: MatPCA and MatFLDA. Patt. Recogn. Lett. 26, 1157–1167 (2005)
5. Eklundh, L., Singh, A.: A comparative analysis of standardized and unstandardized principal component analysis in remote sensing. Int. J. Remote Sens. 14, 1359–1370 (1993)
6. Gao, Q., Zhang, L., Zhang, D., Yang, J.: Comments on On image matrix based feature extraction algorithms. IEEE Trans. Sys. Man Cyber. 37B, 1373–1374 (2007)
7. Guillamet, D., Vitria, J., Schiele, B.: Introducing a weighted non-negative matrix factorization for image classification. Patt. Recogn. Lett. 24, 2447–2454 (2003)
8. Hubert, M., Rousseeuw, P., Verdonck, T.: Robust PCA for skewed data and its outlier map. Comp. Stat. Data Anal. 53, 2264–2274 (2009)
9. Jolliffe, I.T.: Principal Component Analysis, 2nd edn. Springer, Heidelberg (2002)
10. Lee, S.H., Choi, S.: Two-dimensional canonical correlation analysis. IEEE Sig. Process. Lett. 14, 735–738 (2007)
11. Liang, Z., Li, Y., Shi, P.: A note on two-dimensional linear discriminant analysis. Patt. Recogn. Lett. 29, 2122–2128 (2008)
12. Liu, C., Wechsler, H.: Evolutionary pursuit and its application to face recognition. IEEE Trans. Patt. Anal. Mach. Intell. 22, 570–582 (2000)
13. Ma, Y., Ijiri, Y., Lao, S., Kawade, M.: Re-weighting linear discriminant analysis under ranking loss. In: IEEE Int. Conf. CVPR (2008)
14. Meytlis, M., Sirovich, L.: On the dimensionality of face space. IEEE Trans. Patt. Anal. Mach. Intell. 29, 1262–1267 (2007)
15. Nhat, V.D.M., Lee, S.Y.: Image-based subspace analysis for face recognition. In: Delac, K., Grgic, M. (eds.) Face Recognition. I-Tech, Vienna (2007)
16. Park, M.S., Na, J.H., Choi, J.Y.: Feature extraction using class-augmented principal component analysis (CA-PCA). In: Kollias, S.D., Stafylopatis, A., Duch, W., Oja, E. (eds.) ICANN 2006. LNCS, vol. 4132, pp. 606–615. Springer, Heidelberg (2006)
17. Park, M.S., Choi, J.Y.: Theoretical analysis on feature extraction capability of class augmented PCA. Patt. Recogn. 42, 2353–2360 (2009)
18. Price, J.R., Gee, T.F.: Face recognition using direct, weighted linear discriminant analysis and modular subspaces. Patt. Recogn. 38, 209–219 (2005)
19. Qi, Y., Zhang, J.: $(2D)^2$PCALDA: An efficient approach for face recognition. Appl. Maths Comp. (in press, 2009)
20. Qiu, X., Wu, L.: Nearest neighbor discriminant analysis. Int. J. Patt. Recogn. Artif. Intell. 20, 1245–1259 (2006)
21. Qiu, X., Wu, L.: Two-dimensional nearest neighbor discriminant analysis. Neurocomputing 70, 2572–2575 (2007)
22. Sandler, R., Lindenbaum, M.: Nonnegative matrix factorization with earth mover's distance metric. In: Proc. Comp. Vis. Patt. Recog. (2009)
23. Harrison, S.: Standardized principal component analysis. Int. J. Remote Sens. 6, 883–890 (1985)
24. Shah, V.P., Younan, N.H., King, R.L.: An efficient pan-sharpening method via a combined adaptive PCA approach and contourlets. IEEE Trans. Geosci. Remote Sens. 46, 1323–1335 (2008)
25. Turk, M.: Eigenfaces and beyond. In: Zhao, W., Chellappa, R. (eds.) Face processing: advanced modeling and methods. Academic Press, London (2006)

26. Verboven, S., Hubert, M.: LIBRA: a matlab library for robust analysis. Chemometrics Intell. Lab. Sys. 75, 127–136 (2005)
27. Wang, X., Tang, X.: Random sampling for subspace face recognition. Int. J. Comp. Vis. 70, 91–104 (2006)
28. Yang, F., Su, J.: Fast illumination normalization for robust eye localization under variable illumination. J. Elect. Imag. 18, 010503-1–010503-3 (2009)
29. Yang, J., Zhang, D.: Two-dimensional PCA: A new approach to appearance based face representation and recognition. IEEE Trans. Patt. Anal. Mach. Intell. 26, 131–137 (2004)
30. Yang, J., Zhang, D., Yang, J.-Y.: Median fisher discriminator: a robust feature extraction method with applications to biometrics. Frontiers Comp.Sci. China 2, 295–305 (2008)
31. Yu, H., Yang, J.: A direct LDA algorithm for high-dimensional data - with application to face recognition. Patt. Recogn. 34, 2067–2070 (2001)
32. Zhang, D., Chen, S., Zhou, Z.-H.: Two-dimensional non-negative matrix factorization for face representation and recognition. In: Int. Conf. Comp. Vis., pp. 350–363 (2005)
33. Zhang, X., Gao, Y.: Face recognition across pose: A review. Patt. Recogn. 42, 2876–2896 (2009)
34. Zhang, X., Jia, Y.: A linear discriminant analysis framework based on random subspace for face recognition. Patt. Recogn. 40, 2585–2591 (2007)
35. Zhang, D., Zhou, Z.-H., Chen, S.: Diagonal Principal Component Analysis for Face Recognition. Patt. Recogn. 39, 140–142 (2006)

Chapter 10
Ensemble Learning

Abstract. The subject of this chapter is image fusion using the methods of ensemble learning. Ensemble learning is a method for constructing accurate predictors or classifiers from an ensemble of weak predictors or classifiers. In the context of image fusion, we use the term ensemble learning to denote the fusion of K input images $I_k, k \in \{1,2,\ldots,K\}$, where the I_k are all derived from the same base image I^*. The I_k themselves highlight different features in I^*. The theory of ensemble learning suggests that by fusing together the I_k we may obtain a fused image with a substantially improved quality. In the first part of the chapter we consider methods for constructing I_k. In the second part we consider methods for fusing the I_k.

10.1 Ensemble Learning Methods

Ensemble learning is a method for constructing accurate predictors or classifiers from an ensemble of weak predictors or classifiers. Let $\widetilde{E}$ denote the expected classification error of an ensemble of classifiers. Then the theory of ensemble learning [7] suggests that

$$\widetilde{E} = \bar{E} - \bar{D} ,$$

where $\bar{E}$ and $\bar{D}$ are, respectively, the average classification error and diversity of the individual classifiers. This equation discloses that as the accuracy and diversity of an classifiers grows so the performance of an ensemble classifier will also grows.

The following examples illustrate ensemble learning in two different image fusion applications.

Example 10.1. Ensemble Thresholding. In many image processing applications we use a thresholding algorithm to delineate the background from objects of interest or foreground. Unfortunately, in practice, the effectiveness of a given thresholding algorithm is often strongly dependent on the characteristics of the base image I^*. This is illustrated in Fig. 10.1 which shows the results obtained by thresholding three different base images I^* using two different thresholding algorithms. The theory of ensemble learning suggests that by combining several thresholded images B_k we may obtain a thresholded image $\widetilde{B}$ which is less sensitive to the characteristics of the base image I^*.

Example 10.2. Combination of Multiple Edge Operators [3]. Although a large number of diverse edge detection techniques can be found in many image processing publications, there is no single detection method that performs well in every possible image context. Information that could be missed by one detector may be captured by another. Ref. [3] describes a framework for combining multiple edge detection operators in order to yield improved results for edge detection in an image. The so called receiver operating characteristics (ROC) analysis is employed to form an optimum edge map $\widetilde{F}$ that matches the outcomes of a preselected set of edge detectors.

10.2 Diversity Measures

In selecting an ensemble of classifiers we require classifiers which are both accurate and diverse. To estimate the diversity of an ensemble of classifiers we use an appropriate diversity measure. Some common pairwise diversity measures are listed in Table 10.1 [9]. Given an ensemble of K classifiers, the predicted mean ensemble error is

$$\widetilde{E} = (\prod_{k=1}^{K} E_k)^{1/L} \Big(\prod_{h,k,h \neq k} (1 - d_{hk}) \Big)^{1/(L(L-1))} ,$$

where E_k is the mean error rate for the kth classifier and d_{hk} is the pairwise diversity measure for classifiers h and k [6].

Table 10.1 Pairwise Diversity Measures

Name	Description
Yule statistic Q	$Q = (ad - bc)/(ad + bc)$, where a, b, c and d are, respectively, the number of objects that are correctly classified by classifiers S_1 and S_2; are correctly classified by S_1 and incorrectly classified by S_2; are incorrectly classified by S_1 and correctly classified by S_2; are incorrectly classified by S_1 and S_2.
Correlation coefficient ρ	$\rho = (ad - bc)/\sqrt{(a+b)(c+d)(a+c)(b+d)}$.
Disagreement Measure D	$D = (b+c)/N$,where N is the number of training objects.
Double Fault Measure DF	$DF = d/N$.

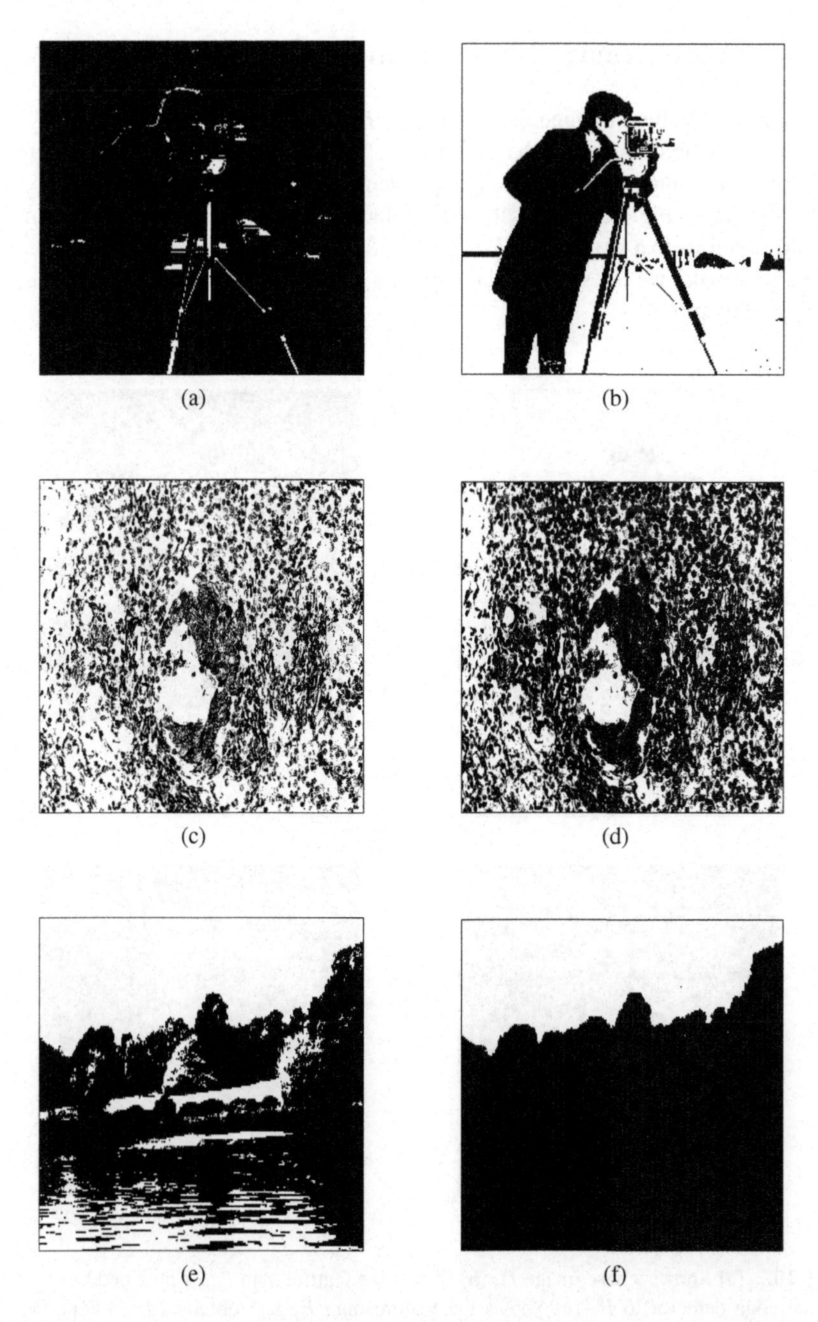

Fig. 10.1 Shows the result of thresholding three images. **(a)**, **(c)** and **(e)** Shows the binary image obtained with the Otsu thresholding algorithm. **(b)**, **(d)** and **(f)** Shows the binary image obtained with the entropy thresholding algorithm. The results show that both algorithms perform reasonably well on "tissue", while the entropy thresholding algorithm fails on "cameraman" and the Otsu thresholding algorithm fails on "autumn".

10.3 Multiple Image Transformations I_k

The simplest method to generate the images $I_k, k \in \{1,2,\ldots,K\}$, is to apply K different transformations to the base image I^*. In order for ensemble learning to be effective, the images I_k should be independent and should highlight different characteristics in I^*. Fig. 10.2 shows the effect of applying three different types of image transformations to a base image I^*.

In Table 10.2 we list some of the common image transformations which may be used for this purpose.

Fig. 10.2 **(a)** Shows a base image I^*. **(b)** Shows the feature map F_{sobel} obtained by applying Sobel edge detector to I^*. **(c)** Shows the feature map $F_{prewitt}$ obtained by a applying the Prewitt edge detector to I^*. **(d)** Shows the feature map F_{zhang} obtained by applying the Zhang-Li wavelet edge detection scheme (see Ex. 8.2) to I^*.

Table 10.2 Image Transformation Techniques

Name	Description
Linear transformation	Transformation in which the important information tends to be contained in the low-frequency coefficients. Examples include: discrete cosine transform, discrete sine transform, and discrete Fourier transform.
Edge operator	Operator which estimates the strength of an edge at all pixels in the image. Examples include: Sobel, Canny, Laplacian-of-Gaussian operator [17].
Directional Filters	Filters which act along specific directions in the image. Examples include Radon and Gabor filters.
Statistical transformation	Change the statistical properties of the image by operating on the image histogram. Examples include histogram matching algorithms.
Multi-scale transformations	Transformation in which the image is decomposed at different scales. Examples include: Laplace and Gaussian pyramid and discrete wavelet transform.

10.3.1 Multiple Subspace Transformations

Subspace transformations are a class of image transforms whose effect is to project I^* into a lower-dimensional space or sub-space. Each projection constitutes a different input image I_k. Further details on the different sub-space transformations is given in Chapt. 9.

One way of generating these sub-spaces is as follows: We perform PCA analysis on a set of K training images. Given the K non-zero eigenvectors $\mathbf{u}_k, k \in \{1,2,\ldots,K\}$, we select α eigenvectors which correspond to the α largest eigenvalues and randomly select β eigenvectors which correspond to the remaining non-zero eigenvalues. The random sub-space is constructed from the $\alpha+\beta$ eigenvectors.

10.3.2 Multiple Random Convolutions

Random convolutions are a class of image transformations which work by convolving the base image I^* with a randomly generated mask M [10]. In [10] the size of the masks are fixed at 3×3 and the mask values are randomly chosen in the range $[-2.5, 2.5]$. Fig. 10.3 shows the effect of convolving a base image I^* with three random convolutions.

$$M_1 = \begin{pmatrix} 1.125 & -0.035 & -0.109 \\ -0.672 & 0.978 & -1.204 \\ 0.267 & 0.655 & 0.804 \end{pmatrix}, \qquad M_2 = \begin{pmatrix} -0.138 & 1.055 & -0.236 \\ 0.289 & 0.596 & 1.089 \\ 0.730 & -0.809 & 1.042 \end{pmatrix},$$

$$M_3 = \begin{pmatrix} -0.224 & -0.368 & -0.903 \\ 0.984 & 0.783 & -0.743 \\ -1.105 & -1.225 & -0.753 \end{pmatrix}.$$

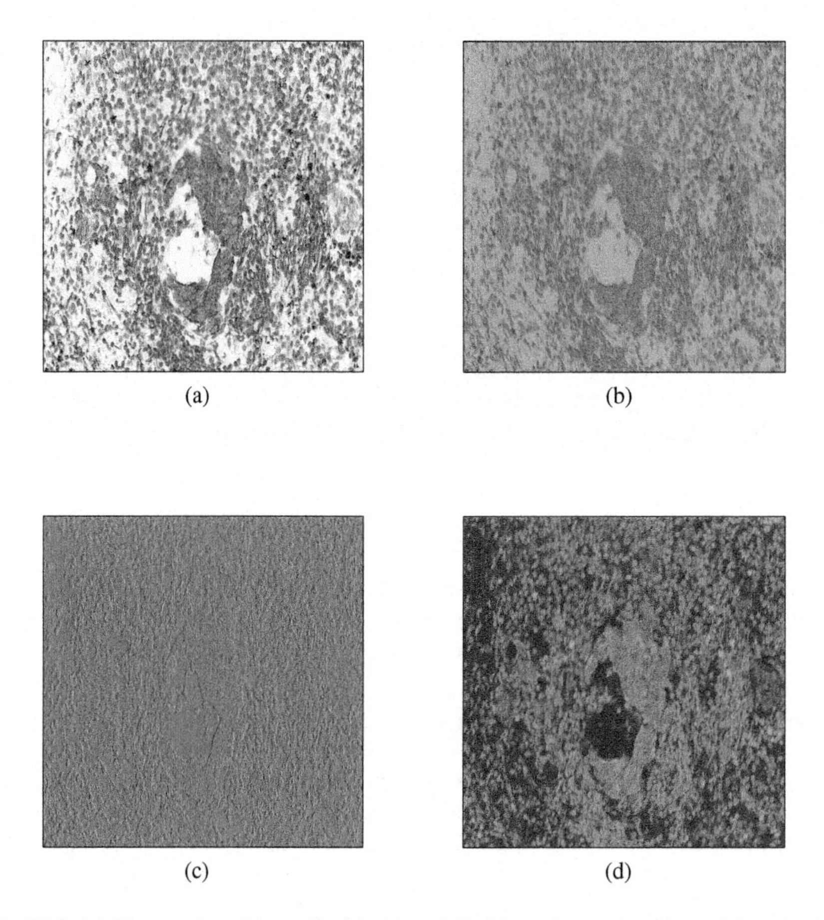

Fig. 10.3 (a) Shows a base image I^*. (b), (c) and (d) Show the result of convolving I^* with the random convolutions M_1, M_2 and M_3.

10.3.3 Multiple Normalizations

Multiple normalizations are used for handwritten character recognition [8]. An input character image is transformed into an ensemble of normalized images using different normalization methods. If R_1 and R_2 are, respectively, the input and output aspect ratios of character image (i. e. the width-to-height ratio), then some common linear normalization methods are:

Fixed aspect ratio

$$R_2 = 1$$

Preserved aspect ratio

$$R_2 = R_1$$

Square root of aspect ratio

$$R_2 = \sqrt{R_1}$$

Cubic root of aspect ratio

$$R_2 = R_1^{1/3}$$

Square root of sine of aspect ratio

$$R_2 = \sqrt{\sin(\pi R_1/2)}$$

By using different normalizations we obtain an ensemble of normalized images I_k.

10.3.4 Multiple Color Spaces

Color space transformations are a class of image transformations whose effect is to transform the base image I^* into a different color space. Recent research has revealed that different color spaces display different discriminating power for pattern recognition. Multiple color spaces is therefore an important method for generating an ensemble of different images I_k. Each color space generates a different input image I_k.

Example 10.3. Fusion of color spaces for ear authentication [12]. The human ear is an important biometric characteristic for the following reasons:

1. The ear structure is rich and full of features.
2. The ear is stable. It changes little with age and facial expression.
3. The ear is relatively easy to capture.
4. The ear is relatively large and thus easily captured at a distance.

The image of the ear is resized to a fixed size using nearest neighbor interpolation. The input images are in *RGB* color space. We transform this image into 12 different color spaces (see Chapt. 16). Features are then extracted from each of the 13 color spaces. Finally the test ear is authenticated using a nearest neighbor classifier (see Ex. 9.3).

Further details on the different color spaces which are available and their corresponding transformations is given in Chapt. 16.

10.3.5 Multiple Thresholds

Image thresholding algorithms are a class of image transformations whose effect is to transform the base image I^* into a set of binary images I_k. Thresholding algorithms may therefore be regarded as classification algorithms in their own right and the images I_k are decision, or label, maps. The thresholding algorithms themselves may be divided into two classes: global algorithms and local algorithms. The global algorithms generate a binary decision map B_k as follows:

$$B_k(x,y) = \begin{cases} 1 \text{ if } I^*(x,y) > t_k , \\ 0 \text{ otherwise} , \end{cases}$$

where $I^*(x,y)$ is the gray-level of the base image at the pixel position (x,y) and t_k is the global threshold value.

Further details on the different global and local thresholding algorithms which are available is given in Chapt. 12. In general we create an ensemble of thresholding algorithms by choosing algorithms which are based on different physical principles. In this case, we are reasonably sure to obtain a diverse ensemble of K binary images $I_k, k \in \{1,2,\ldots,K\}$.

10.3.6 Multiple Segmentations

Image segmentation algorithms are a class of image transformations in which we transform the base image I^* into a set of multi-label images I_k. By choosing segmentation algorithms which are based on different physical principles, we ensure that we obtain a diverse ensemble of K multi-label images $I_k, k \in \{1,2,\ldots,K\}$.

The following example describes a mean shift image segmentation algorithm due to Wang and Suter [19].

Example 10.4. Mean Shift Image Segmentation Algorithm [19]. Let I denote an 8-bit input image with a histogram $H = (H_0, H_1, \ldots, H_{255})^T$, where $H_i, i \in \{0,1,\ldots,255\}$, is the number of pixels in I with a gray-level equal to i. Let P_m and V_m denote, respectively, the mth peak and the mth valley in H, where $V_0 = 0$, $V_M = 255$ and $V_0 \leq P_1 < V_1 \leq \cdots \leq P_M \leq V_M$. We use the following iterative process to find the peaks P_m and the valleys V_m.

1. Initialize $m = 1$.
2. Apply the mean shift peak method to obtain the peak location P_m:
   a. $P_m = V_{m-1} + 1$.
   b. Calculate mean shift vector s
   $$s = \sum_{i=P_m-W}^{P_m+W} iH_i / \sum_{i=P_m-W}^{P_m+W} H_i .$$
   c. $P_m \leftarrow P_m + \alpha(s - P_m)$, where $0 < \alpha < 1$.
   d. Repeat steps (2b)-(2c) until no significant change in s
3. Apply the mean shift valley method to obtain the valley V_m:
   a. $V_m = P_m + 1$.
   b. Calculate mean shift vector s
   $$s = \sum_{i=V_m-W}^{V_m+W} iH_i / \sum_{i=V_m-W}^{V_m+W} H_i .$$

c. $V_m \leftarrow V_m - \alpha(s - V_m)$, where $0 < \alpha < 1$.
d. Repeat steps (3b)-(3c) until no significant change in s
4. Increment $m \leftarrow m + 1$.
5. Repeat steps (2) and (3) until P_m or V_m is equal to, or larger than, 255.
6. Eliminate non-significant peaks. Choose significant valleys as the minimum of the valleys between two consecutive significant peaks.
7. Use the significant valleys to generate a segmented image D with M labels, where

$$D(x,y) = m \qquad \text{if } V_{m-1} \leq I(x,y) < V_m$$

10.4 Re-sampling Methods

Re-sampling methods are a class of special methods which allow for the systematic generation of a virtually unlimited number of I_k in a natural way. Re-sampling methods are considered in detail in chapter 11.

10.5 Image Fusion

Until now we have considered the creation of an ensemble of K images I_k, feature maps F_k or decision maps D_k from a single base image I^*. In this case, the I_k, F_k and D_k are, by definition, spatially and temporally aligned. We shall further assume the I_k, F_k and D_k are semantically equivalent and radiometrically calibrated.

If we restrict ourselves to the pixel-based fusion operators, then for images I_k and feature maps F_k, the most common pixel fusion operators are:

Arithmetic Mean

$$\widetilde{I}(x,y) = \frac{1}{K} \sum_{k=1}^{K} I_k(x,y) .$$

Trimmed Mean

$$\widetilde{I}(x,y) = \frac{1}{K - 2\alpha} \sum_{k=\alpha+1}^{K-\alpha} I_{(k)}(x,y) ,$$

where $I_{(l)}(x,y)$ is the lth largest gray-level at the pixel (x,y) and α is a small integer. Often we let $\alpha \approx \lfloor K/20 \rfloor$.

For decision maps D_k, we cannot use these operators. Instead, we often use a majority-vote or a weighted majority-vote rule:

Majority-Vote

$$\widetilde{D}(x,y) = l \qquad \text{if} \qquad \sum_{k=1}^{K} \delta\left(D_k(x,y), l\right) \geq K/2 , \tag{10.1}$$

where

$$\delta(a,b) = \begin{cases} 1 \text{ if } a = b\,, \\ 0 \text{ otherwise}\,. \end{cases}$$

Weighted Majority-Vote

$$\tilde{D}(x,y) = l \qquad \text{if} \qquad \sum_{k=1}^{K} w_k \delta\left(D_k(x,y),l\right) \geq \sum_{k=1}^{K} w_k/2\,,$$

where each D_k has a weight w_k associated with it.

Additional image fusion methods are considered in Chapts. 7 17 and 21.

In the following example we describe a multi-purpose ensemble-based image classifier [14].

Example 10.5. Multi-Purpose Ensemble Classifier [14]. WND-CHARM is a multi-purpose supervised ensemble classifier. Let $(I_m, y_m), m \in \{1,2,\ldots,M\}$, represent a training set of image/classification pairs, where I_m denotes the mth training image and y_m denotes the corresponding classification. For each image I_m we extract an ensemble of $K = 1025$ feature maps $F_{k,m}$, where $F_{k,m}$ is obtained by applying a feature operator $\hat{F}_k$ to I_m. The feature operators $\hat{F}_k, k \in \{1,2,\ldots,K\}$ are divided into four categories:

Polynomial Decompositions. In polynomial decomposition, a polynomial is generated that approximates the image to some fidelity and the coefficients of the polynomial is used as descriptors of the image content.

Textures. Texture features report on the inter-pixel variation in intensity for several directions and resolutions.

High Contrast features. High contrast features, such as edges and objects, comprise statistics about object number, spatial distribution, size and shape.

Pixel Statistics. Pixel statistics are based on the distribution of pixel intensities within the image and includes histograms and moments.

Due to the high dimensionality of the $\hat{F}_k$, some of the $\hat{F}_k$ are expected to represent noise. WND-CHARM therefore includes an automatic feature selection algorithm. For this purpose we give each operator $\hat{F}_k$ a weight W_k:

$$W_k = \frac{L}{L-1}\frac{\sigma_B^2}{\sigma_W^2}\,,$$

where σ_B^2 is the variance of the class means for feature $\hat{F}_k$ and σ_W is the variance of feature $\hat{F}_k$ among all the training images averaged over all classes l. Mathematically W_k is the Fisher, or LDA, score for F_k (see Chapt. 9). In WND-CHARM 45% of the strongest features are retained which are then used in a nearest neighbour classifier.

The following example illustrates the use of ensemble learning in biometric face recognition.

Example 10.6. Biometric face recognition [5]. Given an input image I^* we create an ensemble of K images $I_k, k \in \{1,2,\ldots,K\}$, by applying K transformations to I^*. In [5] the transformations include histogram equalization (Sect. 6.2.1) and edge detection. The images I_k are then transformed into feature maps F_k by applying a low-dimensional principal component analysis (PCA) transform (Sect. 9.2). Experimental results showed that a significant improvement in facial recognition performance may be obtained when we use an ensemble of images I_k.

We now consider three applications of ensemble fusion.

10.6 Ensemble Thresholding

The effectiveness of a thresholding algorithm is strongly dependent on the input image characteristics. Experimentally we find that for one input image, a given thresholding algorithm may appear the best, while it may fail completely for another image. This makes it difficult to choose the most appropriate algorithm to binarize a given image. One way of solving this problem is to binarize the input image using M different thresholding algorithms and then fusing the resulting binary images together. In this way we exploit the pecularities of the different thresholding algorithms to obtain a more robust final thresholded image. In general, for a given input image, this approach will not outperform the best single thresholding algorithm. However we obtain accuracies which are comparable to those obtained with the best single thresholding algorithm independent of the image statistical characteristics. Given a base image I^* with pixel gray levels $I^*(x,y)$, and K global thresholds $t_k, k \in \{1,2,\ldots,K\}$, we may generate K binary images B_k, where

$$B_k(x,y) = \begin{cases} 1 & I^*(x,y) \geq t_k \, , \\ 0 & \text{otherwise} \, . \end{cases}$$

Two classical fusion strategies which we may use to fuse together the binary images B_k are, respectively, the majority-vote rule and the weighted majority-vote rule. Using the majority-vote fusion rule we obtain a binary image $\widetilde{B}$ whose gray-levels are

$$\widetilde{B}(x,y) = \begin{cases} 1 & \text{if } \sum_{k=1}^{K} B_k(x,y) \geq K/2 \, , \\ 0 & \text{otherwise} \, . \end{cases}$$

Using the weighted majority-vote fusion rule we obtain the binary image

$$\widetilde{B}(x,y) = \begin{cases} 1 & \text{if } \sum_{k=1}^{K} w_k B_k(x,y) \geq \sum_{k=1}^{K} w_k/2 \, , \\ 0 & \text{otherwise} \, , \end{cases} \tag{10.2}$$

where each image B_k has a weight w_k associated with it (Fig. 10.4).

Fig. 10.4 (**a**) Shows the base image I^*. (**b**)-(**f**) Shows binary images $B_k, k \in \{1,2,\ldots,K\}$, obtained using $K=5$ different thresholding algorithms.

(g)

(h)

Fig. 10.4 *cont'd.* **(g)** Shows the binary image $\widetilde{B}$ obtained by fusing the B_k together using a majority-vote rule operator. **(h)** Shows the binary image $\widetilde{B}$ obtained by fusing the B_k together using the weighted majority-vote rule operator (10.2).

Example 10.7. Adaptive Weights [11]. Traditionally, the weights w_k in (10.2) are determined according to the relative performance of each thresholding algorithm. This, however, involves the availability of training data with associated ground-truth. An alternative procedure which does not require any training data is to choose the weights w_k according to how close the kth threshold, t_k, is to the mean threshold $\bar{t} = \sum_k t_k / K$. Mathematically, we let

$$w_k = \exp(-\alpha|t_k - \bar{t}|) . \tag{10.3}$$

where α is a real positive constant which controls the steepness of the weight function. The idea is that the smaller the absolute difference between t_k and $\bar{t}$, the higher the degree of confidence in the classification. Using the same reasoning, we may allow the w_k to vary on a pixel-by-pixel basis. This we do by choosing the weights according to how close the pixel gray-level $I^*(x,y)$ is to the kth threshold t_k. Mathematically, we let

$$w_k(x,y) = 1 - \exp(-\beta|t_k - I^*(x,y)|) ,$$

where β is a real positive constant which controls the steepness of the weight function. The idea is that the larger the difference between the pixel gray-level and the threshold value, the higher the degree of confidence in the classification. For 8-bit deep input images, [11] recommends $\alpha = 0.1 = \beta$.

10.7 Ensemble Spatial Sampling

In ensemble spatial sampling [4] we create a local image thresholding algorithm from a global thresholding algorithm (see Chapt. 12) as follows:. Given an $M \times N$ input image I^* we randomly choose the top-left coordinates (m', n') and the bottom-right coordinates (m'', n'') of a window W. We then threshold the pixels in W using a given global thresholding algorithm. We repeat the entire process K times.

Suppose for the kth iteration, the top-left and bottom-right coordinates are (m'_k, n'_k) and (m''_l, n''_k). The threshold for this window is t_k. The corresponding 3-level thresholded image is:

$$B_k(m,n) = \begin{cases} 1 & \text{if } I^*(m,n) > t^{(k)},\ m'_l \leq m \leq m''_k \text{ and } n''_k \leq n \leq n''_k\ , \\ -1 & \text{if } I^*(m,n) \leq t_k,\ m'_k \leq m \leq m''_k \text{ and } n'_k \leq n \leq n''_k\ , \\ 0 & \text{otherwise}\ . \end{cases}$$

Finally we obtain a fused binary image $\widetilde{B}$ by determining the sign of $\sum_{k=1}^{K} B_k(m,n)$:

$$\widetilde{B}(m,n) = \begin{cases} 1 & \text{if } \sum_{k=1}^{K} B_k(m,n) > 0\ , \\ 0 & \text{otherwise}\ . \end{cases}$$

Fig. 10.5 shows the action of the local thresholding algorithm.

Fig. 10.5 **(a)** Shows the base image I^*. **(b)** Shows the global binary image obtained using the Otsu thresholding algorithm. **(c)** Shows an example of a local 3-level thresholded image B_k. **(d)** Shows the binary image $\widetilde{B}$ obtained by fusing K local 3-level thresholded images $B_k, k \in \{1, 2, \ldots, K\}$.

10.8 Ensemble Atlas Based Segmentation

Atlas-based image segmentation is widely used in medical image applications. The basic idea is explained in the following example.

Example 10.8. Atlas-Based Segmentation [16]. In atlas-based segmentation we have a single training image I^* which we regard as a "gold-standard". From I^* we create the corresponding segmented image D^*. For this purpose we may use any segmentation algorithm to create D^* although it is most common to use a carefully constructed manual segmentation. Given a test image I we spatially align I to I^*. If T is the corresponding spatial transformation:

$$I = T(I^*) ,$$

then we suppose the corresponding segmented image is: $D = T(D^*)$.

In ensemble atlas-based segmentation we generalize this procedure in two different ways:

Multiple Transformations. We suppose we have several different spatial alignment algorithms. Each algorithm gives a different transform $T_m, m \in \{1,2,\ldots,M\}$. By this means we create M segmented images $D_m, m \in \{1,2,\ldots,M\}$, where

$$D_m = T_m(D^*) .$$

Let $\widetilde{D}$ denote the segmented image obtained by fusing the D_m using majority-vote, shape-based average or other appropriate fusion operator f. Mathematically, we have

$$\widetilde{D} = f(D_1, D_2, \ldots, D_M) .$$

Multiple Training Images. We suppose we have several training images $I^*_m, m \in \{1,2,\ldots\}$, each with its own "gold-standard" segmentation D^*_m to generate a different spatial transformations T_m for each training image $I^*_m, m \in \{1,2\ldots,M\}$:

$$I = T_m(I^*_m) .$$

In this case, we use the same spatial alignment algorithm to obtain M spatial transformations $T_m, m \in \{1,2,\ldots,M\}$, one for each I^*_m. By applying T_m to D^*_m, we obtain M segmented images $D_m = T_m(D^*_m)$. As before we may fuse the D_m together using an appropriate fusion operator f:

$$\widetilde{D} = f(D_1, D_2, \ldots, D_M) .$$

10.9 Ensemble Nearest Neighbor Classification

In this section we consider the ensemble nearest neighbor classifier. A nearest neighbor (NN) classifier is defined as follows: Given an unknown object or pattern we classify it by choosing the class of the nearest example in a training set as measured by a given similarity measure.

Given an input image I we may classify it as follows: We create a representative feature vector $\mathbf{f}$:

$$\mathbf{f} = (f_1, f_2, \ldots, f_r)^T .$$

Off-line a small number of these vectors, which we denote as $\mathbf{G}_i, i \in \{1,2,\ldots\}$, have been classified into a set of L classes $c_l, l \in \{1,2,\ldots,L\}$. Then, we compare each r-dimensional vector $\mathbf{f}$ with $\mathbf{G}_i, i \in \{1,2,\ldots\}$. Let $\mathbf{G}^*$ denote the $\mathbf{G}_i$ which is closest (i. e. most similar) to $\mathbf{f}$, then we classify the image I as belonging to the class l^*, where l^* is the class label of $\mathbf{G}^*$. Despite its simplicity, the NN classifier gives a competitive performance with more modern classification techniques.

A simple extension of the NN classifier is to choose the most common class among the K-nearest neighbors. If $S(x,y|l)$ is the number of K nearest neighbors which belong to the lth class, then the K-NN classifier gives the pixel (x,y) the class label l^*, where

$$l^* = \arg\max_l (S(x,y|l)) .$$

Recently Bay [1] described an ensemble nearest neighbor classifier with improved classification performance.

The ensemble NN classifier is constructed as follows. Given the test image I we create M feature maps $F_m(x,y), m \in \{1,2,\ldots,M\}$. This we do by using M transformations, normalizations, directions, color spaces, thresholding or segmentation algorithms. Alternatively, we may simply randomly select s ($s < r$) components from $F(x,y)$. For each F_m we create a decision map $D_m(x,y)$ using a nearest neighbor classifier (see Ex. 9.3). Finally, we fuse the D_m together using the majority-vote rule.

Example 10.9. K-nearest neighbor (K-NN) ensemble classifier [2]. Traditionally we use the majority-vote rule to combine the decision maps $D_m, m \in \{1,2,\ldots,M\}$. However, for the K-NN classifier we may, instead, combine the maps $S_m(x,y|l)$ [2], where $S_m(x,y|l)$ is the number of K nearest samples G_i which belong to the lth class. In this case, the Domeniconci-Yan (DY) fused decision map is

$$\widetilde{D}_{DY}(x,y) = \arg\max_l \Big(\sum_{m=1}^{M} S_m(x,y|1), \sum_{m=1}^{M} S_m(x,y|2), \ldots, \sum_{m=1}^{M} S_m(x,y|L) \Big) .$$

The following is a simple numerical example in which we compare the majority-vote rule and the DY combination rule. Let $M = 3$, $L = 2$ and $K = 5$. Suppose for a given pixel (x,y) we have:

$$S_1(x,y|1) = \frac{2}{5} \quad \text{and} \quad S_1(x,y|2) = \frac{3}{5},$$
$$S_2(x,y|1) = \frac{2}{5} \quad \text{and} \quad S_2(x,y|2) = \frac{3}{5},$$
$$S_3(x,y|1) = \frac{4}{5} \quad \text{and} \quad S_4(x,y|2) = \frac{1}{5}.$$

The individual decision maps $D_m(x,y)$ are

$$D_1(x,y) = 2, \quad D_2(x,y) = 2 \quad \text{and} \quad D_3(x,y) = 1.$$

The corresponding majority-vote and DY fused decision maps are:

$$\widetilde{D}_{maj}(x,y) = \arg\max_l \Big(\sum_{m=1}^{M} D_m(x,y)\Big) = 2,$$
$$\widetilde{D}_{DY}(x,y) = \arg\max_l \Big(\sum_{m=1}^{M} S_m(x,y|l), \sum_{m=1}^{M} S_m(x,y|2), \ldots, \sum_{m=1}^{M} S_m(x,y|L)\Big) = 1.$$

10.10 Further Reading

A modern book devoted to ensemble methods is [13]. For a recent survey of ensemble methods see [18]. For survey of real-world applications using ensemble learning see [15]. For an additional application using ensemble learning see Sects. 16.3.

10.11 Software

STPRTOOL. A statistical pattern recognition toolbox. Authors: Vojtech Franc and Vaclav Hlovac.

References

1. Bay, S.D.: Nearest neighbor classification for multiple feature subsets. Intell. Data Analy. 3, 191–209 (1999)
2. Domeniconci, C., Yan, B.: Nearest neighbor ensemble. In: Proc. 17th Int. Conf. Patt. Recogn. (2004)
3. Giannarou, S., Stathaki, T.: Edge detection using quantitative combination of multiple operators. In: IEEE Workshop on Signal Process. Systems Design and Implement, pp. 359–364 (2005)
4. Hong, Y.: Random spatial sampling and majority voting based image thresholding. IEEE Signal Process. Lett. (2008)

5. Jarillo, G., Pedrycz, W., Reformat, M.: Aggregation of classifiers based on image transformations in biometric face recognition. Mach. Vis. Appl. 19, 125–140 (2008)
6. Ko, A.H.-R., Sabourin, R., de Britto Jr., A.S.: Compound diversity functions for ensemble selection. Int. J. Patt. Recogn. Art. Intell. 23, 659–686 (2009)
7. Krogh Vedelsby (1995)
8. Liu, C.-L., Marukawa, K.: Normalization ensemble for handwritten character recognition. In: Proc. 9th Int. Workshop on Frontiers in Handwriting Recogn. (2004)
9. Kuncheva, L.I.: Combining pattern Classifiers: Methods and Algorithms. John Wiley and Sons, Chichester (2004)
10. Mayo, M.: Random convolution ensembles. In: Ip, H.H.-S., Au, O.C., Leung, H., Sun, M.-T., Ma, W.-Y., Hu, S.-M. (eds.) PCM 2007. LNCS, vol. 4810, pp. 216–225. Springer, Heidelberg (2007)
11. Melgani, F.: Robust image binarization with ensembles of thresholding algorithms. J. Elect. Imaging 15, 023010 (2006)
12. Nanni, L., Lumini, A.: Fusion of color spaces for ear authentication. Patt. Recogn. 42, 1906–1913 (2009)
13. Okum, O., Valentini, G.: Supervised and Unsupervised Ensemble Methods and Their Applications. Springer, Heidelberg (2008)
14. Orlov, N., Shamir, L., Macura, T., Johnston, J., Eckley, D.M., Goldberg, I.G.: WND-CHARM: Multi-purpose image classification using compound image transforms. Patt. Recogn. Lett. 29, 1684–1693 (2008)
15. Oza, N.C., Tumer, K.: Classifier ensembles: select real-world applications. Inf. Fusion 9, 4–20 (2008)
16. Rohlfing, T., Maurer Jr., C.R.: Multiclassifier framework for atlas-based image segmentation. Patt. Recogn. Lett. 26, 2070–2079 (2005)
17. Sharifi, M., Fathy, M., Mahmoudi, M.T.: A classified and comparative study of edge detection algorithms. In: Proc. Int. Conf. Information Technology: Coding and Computing, pp. 117–120 (2002)
18. Rokach, L.: Taxonomy for characterizing ensemble methods in classification tasks: A review and annotated bibliography. Comp. Stat. Data Anal. 53, 4046–4072 (2009)
19. Wang, H., Suter, D.: False-peaks-avoiding mean shift method for unsupervised peak-valley sliding image segmentation. In: Proc 7th Digital Image Computing: Techniques and Applications, pp. 581–590 (2003)

Chapter 11
Re-sampling Methods

Abstract. The subject of this chapter are the re-sampling methods which are a special set of ensemble learning techniques. In the context of image fusion the re-sampling methods create an ensemble of input images $I_k, k \in \{1,2,\ldots,K\}$, from a single base image I^*. In this chapter we shall concentrate on two important re-sampling methods: bootstrapping and boosting.

11.1 Introduction

In the context of image fusion, we use the term ensemble learning to denote the fusion of K input images $I_k, k \in \{1,2,\ldots,K\}$, where the I_k are all derived from the same base image I^*. In the previous chapter we constructed the I_k by applying different signal processing algorithms to I^*. These algorithms include image transformations, normalizations, feature extraction, thresholding and segmentation algorithms. In this chapter we consider a different (re-sampling) approach in which we apply an ensemble of K (fixed) classifiers $C_k, k \in \{1,2,\ldots,K\}$ to I^*. In the re-sampling method the C_k are obtained by training the parameters of a parametric classifier S on re-sampled training data T^*. If D_k denotes the decision map obtained by applying C_k to I^*, then we fuse the $D_k, k \in \{1,2,\ldots,K\}$, together to obtain a fused decision map $\widetilde{D}$. In general, $\widetilde{D}$ is more accurate than any one of the D_k.

The strength of the re-sampling method is that it allows for a systematic generation of a virtually unlimited number of classifiers C_k and the corresponding decision maps D_k in a natural way.

We start with bootstrapping, which is perhaps the most widely used re-sampling method.

11.2 Bootstrapping

Bootstrapping is a basic re-sampling technique in which we generate which an ensemble of K fixed classifiers $C_k, k \in \{1,2,\ldots,K\}$, given a base training data base T^*.

Given T^* we generate K bootstrapped training sets $T_k, k \in \{1,2,\ldots,K\}$, by sampling T^* with replacement (see Ex. 3.10). Then, given a parametric classifier S we create an ensemble of fixed classifiers $C_k, k \in \{1,2,\ldots,K\}$, by separately training S on each training set T_k [1]. If we apply C_k to the test image I^* we obtain a decision map D_k. Finally, we obtain a fused decision map $\widetilde{D}$ by fusing the D_k together:

$$\widetilde{D} = f(D_1, D_2, \ldots, D_K) ,$$

where f is an appropriate fusion operator.

In many cases we use the majority vote rule to fuse the D_k. In this case the combination of bootstrapping and majority vote rule is known as "bagging". In general bagging is useful when the C_k are weak. By this we mean that the performance of the classifier is slightly better than random but is unstable: changes in the training data cause significant changes in the fixed classifier.

We illustrate the concept of bagging on two different applications. The first application illustrates the traditional use of bagging a supervised classifier. The second application illustrates a recent development in which we use bagging in an unsupervised classifier or clustering algorithm.

11.3 Face Recognition with Bagging

We consider face recognition in an unconstrained environment where the appearance of a given face may vary due to changes in lighting, pose and facial expression [5]. In many face recognition applications, only a small number of training samples for each subject are available. These samples are not able to capture all the facial appearance variations. By bootstrapping the training set T^* we generate several subsets $T_k, k \in \{1,2,\ldots,K\}$, of samples from the original training dataset. Each subset T_k is then used to train a classifier C_k.

Given a test face we classify it using the classifiers $C_k, k \in \{1,2,\ldots,K\}$. If D_k is the decision obtained with the classifier C_k, then we obtain a fused decision $\widetilde{D}$ by fusing the D_k together using the majority-vote rule. Blu and Jain [5] found the use of bootstrapping made a substantial improvement in the face recognition accuracy: increasing from 81% to 88.7%.

11.4 Bagged Nearest Neighbor Classifier

The nearest neighbor (NN) classifier (see Sect. 10.9) is simple but very effective classifier which is widely used in many real-world classification systems. It is not, however, a weak classifier, and consequently (conventional) bagging will not improve its performance. However, by creating bootstrapped training sets which are

[1] A parametric classifier is a classifier which has free parameters in it. Optimal values for these parameters are found by training the classifier on a training set. Once the parameter values are specified we have a fixed classifier.

smaller in size than the original training set, we "weaken" the NN classifier which can now be bagged.

Example 11.1. Bagged Nearest Neighbor Classifier [3]. Given a base training set T^* containing M images $y_m, m \in \{1,2,\ldots,M\}$, we create a set of K bootstrapped training sets $T_k, k \in \{1,2,\ldots,K\}$, where each training set T_k contains N images which are selected by randomly sampling T^* with replacement and $N \approx 0.7M$.

Given a test image I we classify it using the NN classifier on each bootstrapped training set $T_k, k \in \{1,2,\ldots,K\}$. Suppose the NN classification obtained with the kth training set is a decision label l:

$$D_k = l \; ,$$

then the bagged NN classification of I is

$$\widetilde{D} = l \qquad \text{if} \qquad \sum_{k=1}^{K} \delta(D_k, l) > \frac{K}{2} \; ,$$

where

$$\delta(a,b) = \begin{cases} 1 \text{ if } a = b \; , \\ 0 \text{ otherwise} \; . \end{cases}$$

11.5 Bagged K-means Clustering

Traditionally, re-sampling methods are used in supervised learning applications in order to improve classification accuracy. Recently re-sampling methods have been used to improve unsupervised clustering algorithms.

In this context we use bagging to generate and aggegrate multiple clusterings and to assess the confidence of cluster assignments for individual observations. The motivation is to reduce the variability in the partitioning results via averaging.

Given an input base image I^* we may segment it into a L label decision image D^* using a K-means clustering algorithm (see Ex. 6.8). Let $G_1, G_2, \ldots, G_K$ denote L cluster centers or cluster gray-levels. Then each pixel gray-level $g_m \equiv I^*(m)$ is assigned to a given cluster:

$$\delta_{ml} = \begin{cases} 1 \text{ if } g_m \text{ is assigned to } G_l \; , \\ 0 \text{ otherwise} \; . \end{cases}$$

We then use the assignment matrix δ_{ml} to create the decision image D^*:

$$D^*(m) = l \qquad \text{if} \qquad \delta_{ml} = 1 \; .$$

The K-means algorithm attempts to find the set of cluster centers $G_l, l \in \{1,2,\ldots,L\}$, such that the total error is a minimum:

$$(G_1,G_2,\ldots,G_L) = \arg\min_{G_k} \sum_{m=1}^{M} \sum_{k=1}^{L} \delta_{mk}|g_m - G_k| ,$$

using the following iterative procedure. In each iteration we calculate the assignment matrix δ_{ml} using the cluster centers G_l calculated in the previous iteration. The cluster centers are then re-calculated using the new assignment matrix. The entire process for T iterations is:

```
for t = 1 : T
    for m = 1 : M
        δ_mk^(t) = { 1 if |g_m − G_l^(t−1)| = min_h |g_m − G_h^(t−1)|
                   { 0 otherwise
    end
    for l = 1 : L
        G_l^(t) = Σ_{m=1}^M δ_ml^(t) g_m / Σ_{m=1}^M δ_ml^(t)
    end
end
```

Dudoit and Fridlyand [1] show how we may improve the performance of the cluster algorithm by bagging.

Example 11.2. Bagged K-means Clustering Algorithm [1]. The steps in the bagged K-means clustering algorithm are:

1. Transform the base image I^* into a column vector $I^*(m), m \in \{1,2,\ldots,M\}$
2. Form K bootstrapped column vectors $I^{(k)}(m), k \in \{1,2,\ldots,K\}, m \in \{1,2,\ldots,M\}$ by sampling I^* with replacement.
3. Train the K-means cluster algorithm on each bootstrapped column vector $I^{(k)}$, i. e. learn L cluster centers $G_l^{(k)}, l \in \{1,2,\ldots,L\}$.
4. For each k, permute the labels l so that $G_1^{(k)} < G_2^{(k)} < \ldots < G_L^{(k)}$. This ensures the semantic equivalence of the labels l (see Ex. 5.5).
5. For each set of cluster centers $G_l^{(k)}, k \in \{1,2,\ldots,K\}$, classify the pixels in I^*:
$$D^{(k)}(m) = \arg\min_l (|I^*(m) - G_l^{(k)}|) .$$
6. For each pixel $m, m \in \{1,2,\ldots,M\}$, form a bagged decision $\widetilde{D}(m)$ using a majority-vote rule:

$$\tilde{D}(m) = l \text{ if } \sum_{k=1}^{K} \delta(D^{(k)}(m),l)) \geq \frac{K}{2} ,$$

where

$$\delta(a,b) = \begin{cases} 1 \text{ if } a = b , \\ 0 \text{ otherwise} . \end{cases}$$

11.6 Boosting

Boosting is closely related to bagging except the training sets $T_k, k \in \{1,2,\ldots,K\}$, are no longer independent, instead they are created sequentially. If C_k denotes the classifier which is trained on T_k, then T_{k+1} is created in accordance with the classification accuracy obtained with C_k.

Adaboosting is probably the most successful boosting algorithm. It creates an ensemble of fixed classifiers $C_k, k \in \{1,2,\ldots,K\}$, as follows. Let $T^*(i), i \in \{1,2,\ldots,N\}$, denote the individual samples in T^*. At the kth iteration, each sample $T^*(i)$ is assigned a weight $w_k(i)$. Together, the training samples $T^*(i)$ and the weights $w_k(i)$ constitute a *weighted* training set T_k. At the kth iteration we create a fixed classifier C_k by training a parametric classifier S on the weighted training set T_k. For the next iteration $(k+1)$ we update $w_k(i), i \in \{1,2,\ldots,N\}$, by increasing the weight of $w_k(i)$ if C_k incorrectly classifies $T^*(i)$ and decreasing the weight of $w_k(i)$ if C_k correctly classifies $T^*(i)$.

The following is the pseudo-code for the two-class adaboost algorithm:

Example 11.3. Adaboost Algorithm

```
Initialize $w_0(i) = 1/N, i \in \{1,2,\ldots,N\}$
For $k = 0 : K$
   Generate $C_k$ by training $S$ on a weighted
       training set $T_k$
   Find samples $T^*(i)$ which are misclassified by $C_k$:
       $\delta_k(i) = \begin{cases} 1 \text{ if } C_k \text{ misclassifies } T^*(i) \\ 0 \text{ otherwise.} \end{cases}$
   Calculate $E_K$ and $\beta_k$:
       $E_k = \sum_{i=1}^{N} \delta_k(i) w_k(i)$, $\beta_k = E_k/(1-E_k)$
   Update the weight vector:
      $w_{k+1}(i) = w_k(i)(\delta_k(i) + (1-\delta_k(i))\beta_k)$
   Normalize $w_{k+1}$: $w_{k+1}(i) = w_{k+1}(i)/\sum_{i=1}^{N} w_{k+1}(i)$
end
```

Given the classifiers $C_k, k \in \{1,2,\ldots,K\}$, we classify a test sample as follows: Let D_k be the classification of the test sample obtained with the classifier C_k. Then we fuse the D_k using a weighted majority-vote rule, where the weight given to each classifier C_k is proportional to its accuracy on the weighted training set used to train C_k. The final classification of the test sample is $\widetilde{D}$:

$$\widetilde{D} = l \qquad \text{if} \qquad \sum_{k=1}^{K} \log(1/\beta_k)\delta(D_k, l) \geq \frac{1}{2}\sum_{k=1}^{K} \log(1/\beta_k) \,.$$

where

$$\delta(a,b) = \begin{cases} 1 \text{ if } a = b \,, \\ 0 \text{ otherwise} \,. \end{cases}$$

Adaboost is widely used in classification applications. Since its introduction in 1997 [2] it has established itself as a high-performance general purpose classifier. However, in noisy situations, such as those with label noise, the adaboost algorithm may perform poorly. In this case we may use the ave2boost algorithm. This is a modified version of adaboost specifically designed for noisy situations. In ave2boost we regulate how the weight vector w_k is updated and we modify the weighted-vote rule. For the sake of completeness we give the pseudo-code for ave2boost:

Example 11.4. Ave2boost Algorithm [8].

```
Initialize w_0(i) = 1/N, i ∈ {1,2,...,N}
For k = 0:K
   Generate C_k by training S on weighted
       training set T_k
   Find samples T*(i) which are misclassified by C_k:
       δ_k(i) = { 1 if C_k misclassifies T*(i)
                { 0 otherwise.
   Calculate E_k, β_k and γ_k:
       E_k = Σ_{i=1}^N δ_k(i) w_k(i), β_k = E_k/(1 − E_k), γ_k = (2(1−E_k)k+1)/(2E_k k+1)
   Update the weight vector:
       w_{k+1}(i) = w_k(i)(δ_k(i) + (1 − δ_k(i))β_k)
   Calculate regularization factors:
       v_{k+1}(i) = w_{k+1}(i)/Σ_{i=1}^N w_{k+1}(i)
   Normalize w_{k+1}: w_{k+1}(i) = (k w_{k+1}(i) + v_{k+1}(i))/(k+1)
end
Classify test sample with each C_k:
    D_k = l if C_k identifies test sample as belonging
         to the lth class
Final classification of test sample is:
    D̃ = l if Σ_{k=1}^K log(1/(β_k γ_k))δ(D_k,l) ≥ ½ Σ_{k=1}^K log(1/(β_k γ_k))
    where δ(D_k,l) = 1 if D_k = l, otherwise δ(D_k,l) = 0.
```

11.7 Viola-Jones Algorithm

The conventional adaboost algorithm uses only one parametric classifier S. Viola and Jones (VJ) [10] remove this restriction and use instead M parametric classifiers $S_m, m \in \{1,2,\ldots,M\}$. In each iteration $k, k \in \{1,2,\ldots,K\}$, we select one fixed classifier C_k as follows:

1. Generate M fixed classifiers $c_{k,m}, m \in \{1,2,\ldots,M\}$, by training $S_m, m \in \{1,2,\ldots,M\}$, on the weighted training set T_k.
2. Select the fixed classifier with the smallest weighted error:

$$C_k = c_{k,m} \qquad \text{if} \qquad m = \arg\min_m \sum_{i=1}^{N} \delta_{k,m}(i) w_k(i) ,$$

where

$$\delta_{k,m}(i) = \begin{cases} 1 \text{ if } c_{k,m} \text{ misclassifies } T^*(i) , \\ 0 \text{ otherwise} . \end{cases}$$

11.8 Boosted Object Detection

The VJ algorithm was originally developed for real-time object detection. In this application the number of parametric classifiers $S_m, m \in \{1,2,\ldots,M\}$, is very large. For example, in typical application concerning face detection in an input image, $M \approx 180000$ and special attention must therefore be paid to generating and efficiently training the S_m.

The boosted object detection algorithm works as follows. We create N training samples $T^*(i), i \in \{1,2,\ldots,N\}$, by dividing several training images into overlapping blocks of size $L \times L$. Each block B_i constitutes a training sample $T^*(i)$, where

$$y(i) = \begin{cases} 1 \text{ if } B_i \text{ contains an object} , \\ 0 \text{ otherwise} . \end{cases}$$

In each iteration k of the algorithm, we probe the blocks $B_i, i \in \{1,2,\ldots,N\}$, with a $L \times L$ ternary mask ϕ_m (Fig. 11.1) and compare the result to a threshold θ:

$$d_{k,m}(i) = \begin{cases} 1 \;\text{ if } p \sum_{(x,y)} \phi_m(x,y) B(x,y) \geq p\theta , \\ 0 \text{ otherwise} . \end{cases}$$

where p is a polarity variable ($p = -1$ or $+1$) which determines if the $(\sum_{(x,y)} \phi_{k,m}(x,y) \times B(x,y))$ should be greater, or smaller, than θ. We fix the polarity variable p and the threshold θ by minimizing the weighted error:

$$e_{k,m} = \sum_{i=1}^{N} w_k(i) |d_{k,m}(i) - y(i)| .$$

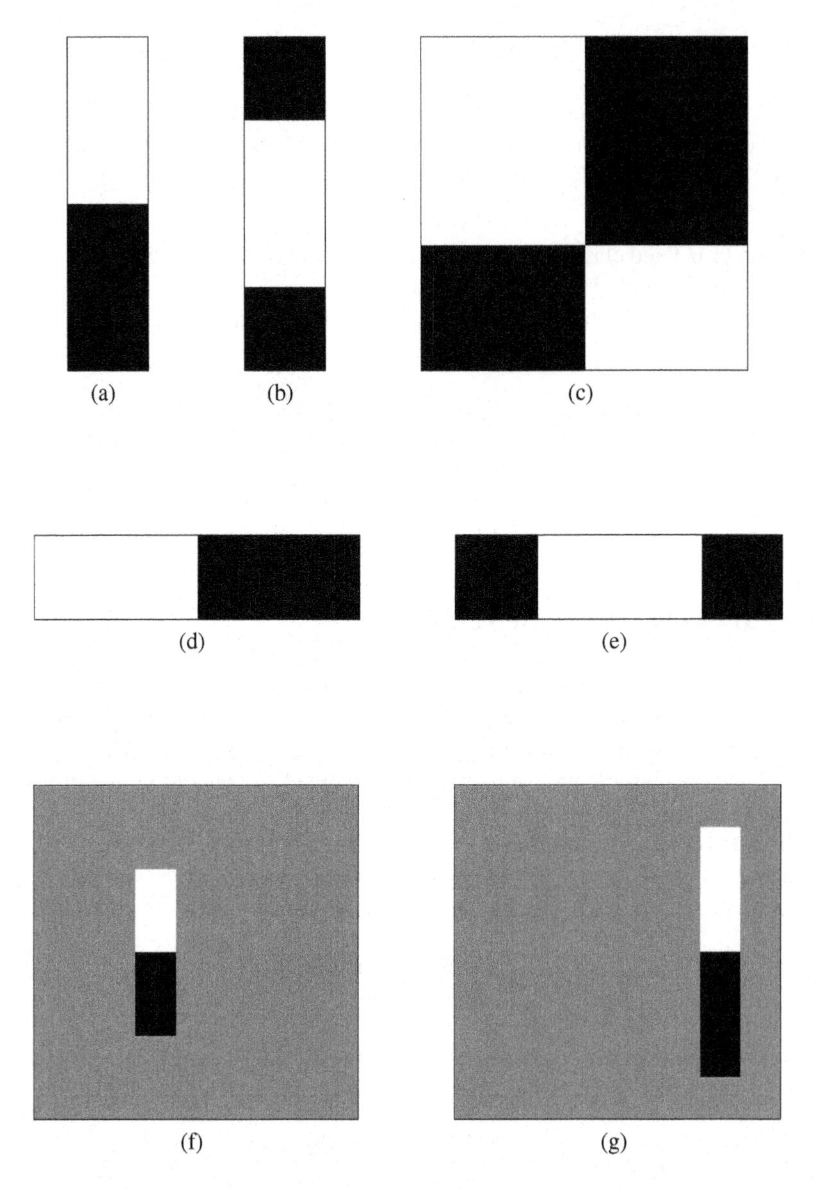

Fig. 11.1 **(a)**-**(e)** Show five basic masks which contain only -1 and $+1$ values. **(f)** Shows a ternary mask ϕ_m obtained by placing the first basic mask in a block of zeros. **(g)** Show a ternary mask ϕ_l obtained by scaling the first basic mask and placing it in a block of zeros. In the figure, white, black, and gray stand, respectively, for values of $+1$, -1 and 0.

The following example is the pseudo-code for the Viola-Jones algorithm.

Example 11.5. Viola-Jones Object Detection Algorithm [6, 10].

1. Extract the features $f_m(i) = \sum_{(x,y)} \phi_m(x,y)B_i(x,y), m \in \{1,2,\ldots,M\}$, for each training block $B(i), i \in \{1,2,\ldots,N\}$.
2. Initialize the weight $w_1(i)$ for each training block i:

$$w_1(i) = \begin{cases} 1/(2N_{true}) & \text{if } B(i) \text{ contains an object} , \\ 1/(2N_{false}) & \text{if } B(i) \text{ does not contain an object} , \end{cases}$$

   where N_{true} is the number of training blocks which contain an object and N_{false} is the number of training blocks which do not contain an object.
3. For $k = 1 : K$ perform the following:
   a. Normalize the weights $w_k(i)$:

$$w_{k+1}(i) = w_k(i) / \sum_{i=1}^{N} w_{k+1}(i) .$$

   b. For each features f_m create a fixed classifier $c_{k,m}$ by training S_m on the weighted error:

$$e_{k,m} = \sum_{i=1}^{N} w_k(i) |d_{k,m}(i) - y(i)| ,$$

   c. Feature selection. Choose the fixed classifier C_k with the lowest weighted error E_k:

$$\left. \begin{array}{l} C_k = c_{k,m} \\ E_k = e_{k,m} \end{array} \right\} \quad \text{if} \qquad m = \arg\min_n (e_{k,n}) ,$$

   d. Update the weights $w_k(i)$:

$$w_{k+1}(i) = w_k(i) \beta_k^{(1-\Delta_k(i))} .$$

   where $\beta_k = E_k/(1-E_k)$ and

$$\Delta_k(i) = \begin{cases} 1 \text{ if } C_k \text{ incorrectly classifies } B(i) , \\ 0 \text{ otherwise} . \end{cases}$$

4. For a test block B we separately classify it using the fixed classifiers $C_k, k \in \{1,2,\ldots,K\}$. If $D_k, k \in \{1,2,\ldots,K\}$, are the corresponding decisions, then we combine the D_k to obtain a fused decision $\tilde{D}$:

$$\tilde{D} = \begin{cases} 1 \text{ if } \sum_{k=1}^{K} \alpha_k D_k \geq \sum_{k=1}^{K} \alpha_k/2 , \\ 0 \text{ otherwise} , \end{cases}$$

   where $\alpha_k = \log(1/\beta_k)$.

For real-time processing, the parametric classifiers S_m are constructed by scaling a basic binary mask (containing only the values -1 and $+1$) and placing it anywhere in a $L \times L$ background of zeros. This method of construction facilitates real-time processing by using method of integral images which we now explain in the following example.

Example 11.6. Integral Image. Given an $L \times L$ image block B with pixel gray-levels $B(x,y), x,y \in \{1,2,\ldots,L\}$, we pre-compute the following sum:

$$I_\Sigma(x,y) = \sum_{u=1}^{x} \sum_{v=1}^{y} B(u,v) .$$

Then the feature value $f(B) = \sum_{(x,y)} B(x,y)\phi(x,y)$, corresponding to the mask ϕ shown in Fig. 11.2 may be efficiently calculated in four operations:

$$f(B) = I_\Sigma(x_2,y_2) - I_\Sigma(x_1,y_1) - (I_\Sigma(x_4,y_4) - I_\Sigma(x_3,y_3)) .$$

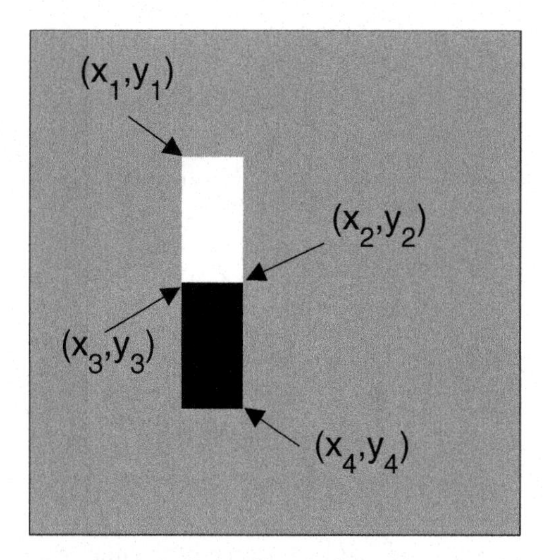

Fig. 11.2 Shows a ternary mask ϕ obtained by placing a binary mask in a block of zeros. In the figure white, black and gray stand, respectively, for values of $+1$, -1 and 0. The top left-hand corner and bottom right-hand corner of the $+1$ block are (x_1,y_1) and (x_2,y_2). Similarly, the top left-hand corner and bottom right-hand corner of the -1 block (x_3,y_3) and (x_4,y_4).

11.9 Software

STPRTOOL. A statistical pattern recognition toolbox. Authors: Vojtech Franc and Vaclav Hlovac.

11.10 Further Reading

A modern book devoted to ensemble methods is [7]. Ref. [4] is a detailed report on implementation of the Viola-Jones algorithm for real-time object detection. For a recent extension of the Viola-Jones algorithm for object detection see [9]

References

1. Dudoit, S., Fridlyand, J.: Bagging to improve the accuracy of a clustering procedure. Bioinformatics 19, 1090–1099 (2003)
2. Freund, Y., Schapire, R.E.: Decision-theoretic generalization of on-line learning and an application to boosting. J. Comp. System Sci. 55, 119–139 (1997)
3. Hall, P., Samworth, R.J.: Properties of bagged nearest neighbor classifiers. J. R. Statist. Soc. 67B, 363–379 (2005)
4. Jensen, O.H.: Implementing the Viola-Jones face detection algorithm. MSc thesis, Technical University of Denmark (2008)
5. Lu, X., Jain, A.K.: Resampling for Face Recognition. In: Kittler, J., Nixon, M.S. (eds.) AVBPA 2003. LNCS, vol. 2688. Springer, Heidelberg (2003)
6. Martins, R., Pina, P., Margques, J.S., Silveira, M.: Crater detection by a boosting approach. IEEE Geosci. Remote Sensing Lett. 6, 127–131 (2009)
7. Okum, O., Valentini, G.: Supervised and Unsupervised Ensemble Methods and Their Applications. Springer, Heidelberg (2008)
8. Oza, N.C.: Ave2boost: boosting for noisy data. In: Roli, F., Kittler, J., Windeatt, T. (eds.) MCS 2004. LNCS, vol. 3077, pp. 31–40. Springer, Heidelberg (2004)
9. Pavani, S.-K., Delgado, D., Frangi, A.F.: Haar-like features with optimally weighted rectangles for rapid object detection. Patt. Recogn. (in print, 2009)
10. Viola, P., Jones, M.: Robust real-time face detection. Int J. Comp. Vis. 57, 137–154 (2004)

Chapter 12
Image Thresholding

Abstract. The subject of this chapter is image thresholding in which we transform an input image, A, into a binary image B, where the pixel gray-levels in B are restricted to $\{0,1\}$. If a_m is the gray level of the mth pixel in A, then the corresponding value in B is

$$b_m = \begin{cases} 1 \text{ if } a_m \geq t_m \ , \\ 0 \text{ otherwise} \ , \end{cases}$$

where t_m is the threshold value for the mth pixel. The thresholds $t_m, m \in \{1,2,\ldots,M\}$, may all be equal to a global threshold t_G or they may vary locally (i. e. from pixel to pixel). In this chapter we shall concentrate on unsupervised thresholding methods. These are thresholding algorithms in which we only use information contained in the current input image to calculate t_m and t_G.

12.1 Global Thresholding

Let A denote a given input image, where a_m denotes the gray-level of the mth pixel. Then the pixels in the image are divided into two groups: "low" intensity pixels whose gray-levels are less than, or equal to, a threshold t and "high" intensity pixels whose gray-levels are greater than t. We follow [12] and categorize the thresholding methods into five groups as follows:

Histogram-based Methods. These methods analyze the shape and features of the image histogram.

Cluster-based Methods. These methods cluster the pixel gray-levels into two parts as background and foreground pixels.

Entropy-based Methods. The methods use the entropy of the histogram or cross-entropy between input image and the thresholded image to find an optimal threshold.

Object Attribute-based methods. The methods use a measure of similarity between the gray-level and the binarized images.

Spatial-based Methods. The methods use a higher-order probability distribution to model the correlation between pixels.

In some applications we find it useful to modify the input image before thresholding. The following example describes one such application involving the estimation of raindrop sizes.

Example 12.1. Histogram Modification [13]. Rainfall estimates often require an estimate of the distribution of raindrop sizes. Unfortunately image irregularities mean the size estimates are often noisy and inaccurate. In [13] the image irregularities are reduced by applying the following iterative histogram modification algorithm to the raindrop images.

The algorithm iteratively sharpens the peaks of the input image histogram by considering the number of pixels H_i having a particular gray-level i and comparing it with $\bar{H}_i$:

$$\bar{H}_i = \frac{1}{2R}\left(\sum_{k=i-R}^{i-1} H_i + \sum_{k=i+1}^{i+R} H_i\right) ,$$

where $\bar{H}_i$ is the average number of pixels in the neighboring R bins on either side of i. Whenever $H_i > \bar{H}_i$, we shift $\lfloor xH_l \rfloor$ pixels from the lth bin to the neighboring bin nearest to the ith bin, where

$$x = \frac{H_i - \bar{H}_i}{H_i} .$$

12.2 Statistical Algorithms

Many of the global thresholding algorithms are statistical in nature. In this case, the threshold t is found by analyzing the image histogram $\mathbf{H} = (H_0, H_1, \ldots, H_{K-1})^T$, or equivalently, the probability distribution $\mathbf{p} = (p_0, p_2, \ldots, p_{K-1})^T$, where H_k is the number of pixels in the input image whose gray levels are equal to k [1] and $p_k = H_k / \sum_{l=0}^{K-1} H_l$.

Fig. 12.1 shows binary images $B_k, k \in \{1, 2, \ldots, K\}$, obtained by thresholding an input image I using K different global thresholding algorithms.

[1] For an 8-bit deep input image the gray-levels are $0, 1, \ldots, 255$, i. e. $K = 256$.

Fig. 12.1 **(a)** Shows an input image I. **(b)**-**(e)** Shows the binary image obtained by thresholding I using the Otsu, Kittler-Illingsworth, Kapur and Tsai thresholding algorithms. **(f)** Shows the binary image obtained by thresholding I using a maximum likelihood algorithm in which we assume a mixture of two Gaussian distributions [1].

12.2.1 Ridler-Calvard

Ridler and Calvard [10] describe an iterative heuristic thresholding technique. The initial threshold, t, is set equal to the mean pixel gray-level $\bar{\mu}$:

$$\bar{\mu} = \sum_{k=0}^{K} kH_k / \sum_{k=0}^{K} H_k .$$

Thereafter, the threshold value t_k for the kth iteration is given by:

$$t_k = \frac{\mu_0(t_k) + \mu_1(t_k)}{2} ,$$

where $\mu_0(t_k)$ and $\mu_1(t_k)$ are, respectively, the mean gray-levels of the pixels which lie below and above the threshold t_k.

12.2.2 Otsu

According to Otsu [9] we select a threshold t which minimizes the within-group variance $\sigma_W^2(t)$ of the pixels. Mathematically, $\sigma_W^2(t)$ is defined as follows:

$$\sigma_W^2(t) = P_0(t)\sigma_0^2(t) + P_1(t)\sigma_1^2(t) ,$$

where $P_0(t)$ and $P_1(t)$ are, respectively, the proportion of pixel gray-levels which lie below, and above, the threshold t and $\sigma_0^2(t)$ and $\sigma_1^2(t)$ are, respectively, the variance of the pixel gray-levels which lie below, and above, the threshold t. The optimum threshold may be found by an exhaustive search. In Ex. 6.6 we illustrate the action of the Otsu algorithm to segment an input image into areas of background and areas of foreground.

For an efficient method of implementating the Otsu algorithm see [4].

12.2.3 Kittler-Illingworth

According to Kittler and Illingworth [7] we select a threshold t which minimizes a criterion $J(t)$, where

$$J(t) = 1 + 2(P_0(t) \ln \sigma_0(t) + P_1 \ln \sigma_1(t)) - 2(P_0(t) \ln P_0(t) + P_1(t) \ln P_1(t)) .$$

where $P_0(t)$ and $P_1(t)$ are, respectively, the proportion of pixel gray-levels which lie below and above the threshold t. If the below threshold and the above threshold pixels each follow a Gaussian distribution, then the criterion $J(t)$ represents the average pixel classification error rate. As the threshold t is varied, the parameters $P_0(t), P_1(t), \sigma_0^2(t)$ and $\sigma_1^2(t)$ change. The problem of minimum error threshold selection is reduced to computing $J(t)$ for all t, and finding its minimum value.

Note. Even if the assumption of a bimodal normal distribution is valid , the model parameters are biased estimates of the true values, as the tails of the overlapping distributions are truncated. Thus, the correctness of the estimated threshold relies on this overlap being small. However, the most challenging images are those where the histograms do not have two well separated modes [3].

12.2.4 Kapur

According to Kapur [6] we select a threshold t which maximizes the sum of the entropies of pixel gray-levels which lie below the threshold t and which lie above the threshold t. Mathematically, the sum of the entropies is

$$\psi(t) = \ln(P_0(t)P_1(t)) + \frac{H_t}{P_0(t)} + \frac{H_G - H_t}{P_1(t)} ,$$

where

$$\begin{aligned} H_t &= -\sum_{k=0}^{t} p_k \ln p_k , \\ H_G &= -\sum_{k=0}^{K-1} p_k \ln p_k . \end{aligned}$$

The discrete value of t which maximizes $\psi(t)$ is the threshold value which maximizes the information between object and background.

12.2.5 Tsai

According to Tsai [14] we select a threshold t such that the binary image has the same first three moments as the input image, where the ith moment of the image I is

$$m_i = \sum_{k=0}^{K-1} k^i p_k ,$$

and, by definition, $m_0 = 1$.

Example 12.2. Change Detection in Remotely Sensed Imagery [8]. The most common methodology to carry out an unsupervised change detection in remotely sensed imagery is to compare two spatially aligned multitemporal remote sensing images $I_1(m,n)$ and $I_2(m,n)$ taken at two different dates over the same geographical area. The result of the comparison is a binary image $B(m,n)$ where

$$B(m,n) = \begin{cases} 1 & \text{if the pixel } (m,n) \text{ is classified as "changed"} , \\ 0 & \text{otherwise} . \end{cases} \quad (12.1)$$

One way of generating B is to threshold the difference image $D = |I_1 - I_2|$ using one of the above global thresholding algorithms. A more powerful method is to match the local gray-level distributions in the two images: For each pixel (m,n) we establish a window W of size $(2L+1) \times (2L+1)$ centered on (m,n). Let H_1 and H_2 denote, respectively, the local distributions, or histograms, of the pixel gray-levels $I_1(i,j)$ and $I_2(i,j), (i,j) \in W$. We then compare H_1 and H_2 using an appropriate similarity measure (see Chapt. 14). We often use the mutual information $MI(H_1,H_2)$ or the Kullback-Leibler distance $KL(H_1,H_2)$ for this purpose.

For a method of combining multiple thresholded images using an *unsupervised* weighted majority vote rule see Ex. 10.7.

12.3 Local Thresholding

In general local thresholding algorithms work by moving a sliding window over the input image. In each window we calculate a threshold t using one of the above global thresholding algorithms. Then t is applied to the center pixel of the window.

However this approach suffers from the drawback that there is no principled method for choosing the local window size. A recent alternative method is the following "ensemble-learning" algorithm (see Sect. 10.7) which may be used to convert the global thresholding algorithms into local algorithms.

Given an input image I we randomly select K image patches from I. The size and location of the image patches is random and the patches may overlap. Thus a given pixel (m,n) may be present in L image patches, where $0 \leq L \leq K$. We threshold each image patch using a global thresholding algorithm. Then for each pixel (m,n) we calculate the number of times it has a label of one. If the number of times is greater than, or equal to, $L/2$, then we set $B(m,n)$ equal to one, otherwise $B(x,y) = 0$.

12.4 Software

HISTHRESH. HistThresh is a matlab toolbox for global image thresholding. Author is Antti Niemisto.

12.5 Further Reading

Recently the expectation-maximization (EM) algorithm has been used for image thresholding assuming the foreground and background pixels follow a given distribution. In [1] the foreground and background pixels are assumed to follow a generalized Gaussian distribution. A comprehensive survey of image thresholding methods is [12]. A survey of entropy and relative entropy thresholding methods is [2]. Two earlier reviews are: [5] and [11].

References

1. Bazi, Y., Bruzzone, L., Melgani, F.: Image thresholding based on the EM algorithm and the generalized Gaussian distribution. Patt. Recogn. 40, 619–634 (2007)
2. Chang, C.-I., Du, Y., Wang, J., Guo, S.-M., Thouin, P.D.: Survey and comparative analysis of entropy and relative entropy thresholding techniques. IEE Proc. Vis. Image Sig. Process. 153, 837–850 (2006)
3. Cho, S., Haralick, R., Yi, S.: Improvement of Kittler and Illingworth's minimum error thresholding. Patt. Recogn. 22, 609–618 (1989)
4. Dong, L., Yu, G., Ogunbona, P., Li, W.: An efficient iterative algorithm for image thresholding. Patt. Recogn. Lett. 29, 1311–1316 (2008)
5. Glasbery, C.A.: An analysis of histogram-based thresholding algorithms. Comp. Vis. Graphics Image Process. 55, 532–537 (1993)
6. Kapur, J.N., Sahoo, P.K., Wong, A.K.C.: A New Method for Gray-Level Picture Thresholding Using the Entropy of the Histogram. Comp. Vis. Graphics Image Process. 29, 273–285 (1985)
7. Kittler, J., Illingworth, J.: Minimum error thresholding. Patt. Recogn. 19, 41–47 (1986)
8. Melgani, F., Bazi, Y.: Markovian fusion approach to robust unsupervised change detection in remotely sensed imagery. IEEE Geosci. Remote Sensing Lett. 3, 457–461 (2006)
9. Otsu, N.: A threshold selection method from gray-level histogram. IEEE Trans. Syst. Man Cyber. 9, 62–66 (1979)
10. Ridler, T.W., Calvard, S.: Picture thresholding using an interative selection method. IEEE Trans. Syst. Man Cyber. 8, 630–632 (1978)
11. Sahoo, P.H., Soltani, S., Wong, A.K.C.: A survey of thresholding techniques. Comp. Vis. Graphics Image Process. 41, 233–260 (1988)
12. Sezgin, M., Sankur, B.: Survey over image thresholding techniques and quantitative performance evaluation. J. Elect. Imag. 13, 146–165 (2004)
13. Sivasubramanian, N.A., Saylor, J.R.: Application of a histogram modification algorithm to the processing of raindrop images. Opt. Engng. 47, 037011-1–037011-10 (2008)
14. Tasi, W.: Moment-preserving thresholding: a new approach. Comp. Vis. Graphics Image Process. 29, 377–393 (1985)

Chapter 13
Image Key Points

Abstract. The subject of this chapter is image key points which we define as a distinctive point in an input image which is invariant to rotation, scale and distortion. In practice, the key points are not perfectly invariant but they are a good approximation. To make our discussion more concrete we shall concentrate on two key point algorithms: SIFT and SURF and their use in spatial alignment.

13.1 Scale-Invariant Feature Transform

The scale-invariant feature transform [5] (SIFT) algorithm provides a robust method for extracting distinctive features from an input image I which are invariant to rotation, scale and distortion. These points (known as "key-points") are found by detecting local extrema in a multi-scale representation of I:

$$\{D(m,n|\sigma_1), D(m,n|\sigma_2), \ldots, D(m,n|\sigma_K)\} , \qquad (13.1)$$

where $D(m,n|\sigma_k)$ is the difference-of-Gaussian (DoG) representation of I at a scale σ_k:

$$D(m,n|\sigma_k) = I(m,n) \otimes G(\sigma_k) - I(m,n) \otimes G(\sigma_{k+1}) ,$$

and $I(m,n) \otimes G(\sigma_k)$ denotes the convolution of $I(m,n)$ with the two-dimensional zero-mean Gaussian $G(\sigma_k)$ and $\sigma_{k+1} = 2^{1/3}\sigma_k$. The parameter K is specified by the user and is based on the maximum width in pixels.

The local extrema in (13.1) are defined as points $(m,n|\sigma_k)$ for which $D(m,n|\sigma_k)$ is greater than its 26 neighbors. This includes eight immediate neighbors from the $D(p,q|\sigma_k)$ and nine neighbors from $D(p,q|\sigma_{k+1})$ and nine neighbors from $D(p,q|\sigma_{k-1})$. This is followed by accurate interpolation of scale space using the Taylor series expansion upto a second degree of $D(m,n|\sigma_k)$ in the neighborhood of (m,n) and σ_k.

Stability of the extrema is further ensured by rejecting key-points with low contrast and key points localized along edges. For a descriptor of the key-point, an orientation histogram is computed of the area surrounding the key-point. Gradient magnitude and the weight of a Gaussian window originating at the key-point add to the value of each sample point within the considered region.

Mathematically the SIFT operator is computed by partitioning the image region surrounding each detection key point into a 4×4 grid of subregions, and computing an orientation histogram of 8 bins in each sub-region (Fig. 13.1).

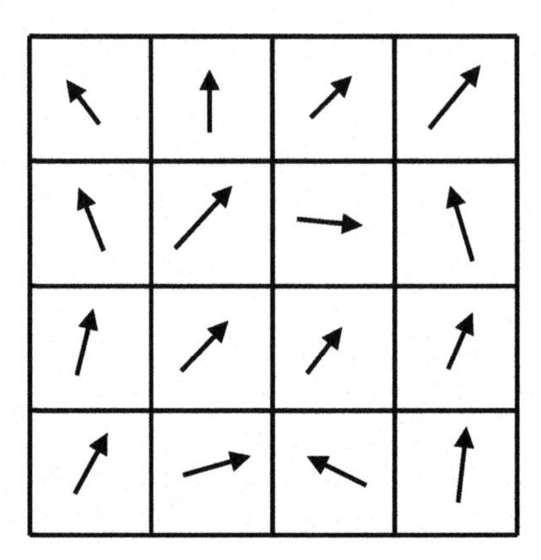

Fig. 13.1 Shows the formation of the SIFT descriptor for a key-point located at (m,n). In the figure we show the 4×4 sub-regions with their orientation vectors.

The grid is square, with the x-axis oriented along the key-point gradient direction and the width of the grid being approximately 12 times the detected scale of the key-point. Within each subregion, the gradient orientation of each pixel is entered into the orientation histogram, with weighted vote proportional to the gradient magnitude. A normalized 128 component vector is formed by concatenating the 16 region containers.

13.1.1 Hyperspectral Images

The SIFT operator has been extended to color images [1, 3] and hyperspectral images [6] as follows. Given a hyperspectral image

$$I(m,n|l), l \in \{1,2,\ldots,L\} ,$$

with L bands, we separately perform the DoG operation on each band [6]:

$$D(m,n|l) = I(m,n|l) \otimes G(\sigma_k) - I(m,n|l) \otimes G(\sigma_{k+1}) ,$$

and then combine the $D_l(m,n|l)$ using a non-linear function f:

$$\widetilde{D}(m,n) = f\left(D(m,n|1), D(m,n|2), \ldots, D(m,n|L)\right) .$$

The function $\widetilde{D}(m,n)$ is then processed in the same way as $D(m,n)$ is processed in the conventional SIFT operator. In hyperspectral images the number of bands L may be one or two hundred. In this case, we may reduce the computational load by reducing the number of bands by applying a PCA operator [6].

For the non-linear function f we may use several different alternatives. Two simple functions are:

$$\widetilde{D}(m,n) = \begin{cases} \sum_{l=1}^{L} |D(m,n|l)|/L , \\ \sum_{l=1}^{L} w_l |D(m,n|l)| / \sum_{l=1}^{L} w_l , \end{cases}$$

where

$$w_l = \begin{cases} 1 \text{ if } D(m,n|l) > \text{median}\left(D(m,n|1), D(m,n|2), \ldots, D(m,n|L)\right) , \\ 0 \text{ otherwise} . \end{cases}$$

13.2 Speeded-Up Robust Feature

The speeded-up robust feature (SURF) [2] algorithm is a variation of the SIFT algorithm. Its major difference includes using a Hessian matrix:

$$H(x,y) = \begin{pmatrix} I_{xx}(x,y|\sigma) & I_{xy}(x,y|\sigma) \\ I_{xy}(x,y|\sigma) & I_{yy}(x,y|\sigma) \end{pmatrix} ,$$

as an interest point detector, where $I_{xx}(x,y|\sigma)$ is the convolution of the Gaussian second-order derivative $\partial^2 G(\sigma)/\partial x^2$ with the input image at (x,y) and similarly for $I_{xy}(x,y|\sigma)$ and $I_{yy}(x,y|\sigma)$. SURF uses integral images (see Ex. 11.6) to calculate $H(x,y)$. This makes SURF much faster than SIFT but with comparable performance.

13.3 Complex Wavelet Transform

The dual-tree complex wavelet transform (see Chapt. 8) provides us with a radically different type of key-point detector [4].

Example 13.1. Dual-Tree Complex Wavelet Transform for Key-Point Detection [4]. The dual-tree CWT decomposition of an $M \times N$ input image I results in a decimated dyadic decomposition into L levels, where each level l is a complex image of size $M/2^l \times N/2^l$. Thus at each (decimated) location (i,j) in level l, we have a set of six complex coefficients: $y_l^{(m)}(i,j) \equiv \rho_l^{(m)} e^{i\theta_m}, m \in \{1,2,\ldots,6\}$, corresponding to the $M = 6$ orientations $\theta_m = \{15^\circ, 45^\circ, 75^\circ, 105^\circ, 135^\circ, 165^\circ\}$. We may define a key-point energy measure $E_l(i,j)$ as

$$E_l(i,j) = \alpha^l \left(\prod_{m=1}^{M} \rho_l^{(m)}\right)^{\beta},$$

where α and β are two parameters whose recommended values are: $\alpha = 1$ and $\beta = 0.25$. We then use the $E_l(i,j)$ values as indicating the presence, or otherwise, of a key-point at level l and decimated location (i,j).

13.4 Software

SIFT. Matlab routine for SIFT operator. Author: Andrea Vedaldi.

References

1. Abdel-Hakim, A.E., Farag, A.A.: CSIFT: A sift descriptor with color invariant characteristics. In: Proc. IEEE Conf. Comput. Vis. Pattern Recog., vol. 2, pp. 1978–1983 (2006)
2. Bay, H., Tuytelaars, T., Van Gool, L.: SURF: Speeded up robust features. In: Leonardis, A., Bischof, H., Pinz, A. (eds.) ECCV 2006. LNCS, vol. 3951, pp. 404–417. Springer, Heidelberg (2006)
3. Brown, M., Lowe, D.: Invariant features from interest point groups. In: Proc. Brit. Mach. Vis. Conf., pp. 656–665 (2002)
4. Fauqueur, J., Kingsbury, N., Anderson, R.: Multiscale key-point detection using the dual-tree complex wavelet transform. In: Proc. Int. Conf. Image Process. (2006)
5. Lowe, D.G.: Distinctive image features from scale-invariant key-points. Int. J. Comput. Vis. 60, 91–100 (2004)
6. Mukherjee, A., Velez-Reyes, M., Roysam, B.: Interest points for hysperspectral image data. IEEE Trans. Geosci. Remote Sensing 47, 748–760 (2009)

Chapter 14
Image Similarity Measures

Abstract. The subject of this chapter is image similarity measures. These measure provide a quantitative measure of the degree of match between two images, or image patches, A and B. Image similarity measures play an important role in many image fusion algorithms and applications including retrieval, classification, change detection, quality evaluation and registration. For the sake of concreteness we shall concentrate on intensity based similarity measures.

14.1 Introduction

Comparing two input images, or image patches, is a fundamental operation in many image fusion algorithms [21, 25, 26]. A meaningful image similarity measure [1] has two components: (1) A transformation T. This extracts the characteristics of an input image and represents it as multi-dimensional feature vector. (2) A distance measure D. This quantifies the similarity between the two images, where D is defined in the multi-dimensional feature space.

Mathematically, we represent a similarity measure between two images A and B as

$$S(A,B) = D(T(A),T(B)) \ .$$

The following example illustrates the use of a similarity measure in a content-based image retrieval (CBIR) system.

Example 14.1. A CBIR System [7, 13]. A CBIR system aims to recover images from an image repository or database, according to the user's interest. In the CBIR system each image in the database is represented as a multi-dimensional

[1] We use the term "similarity measure" as a general term which includes both similarity measures (which reach their maximum value when $A = B$) and dissimilarity, or distance, measures (which reach their minimum value when $A = B$). Apart from mutual information MI all the measures discussed in this chapter are dissimilarity measures.

feature vector which is extracted from a series of low-level descriptors, such as a color histogram, a co-occurrence matrix, morphological features, wavelet-based descriptors or Zernike moments. The subjective similarity between two pictures is quantified in terms of a distance measure which is defined on the corresponding multi-dimensional feature space. Common distance measures are: the Minkowski distance, the Manhattan distance, the Euclidean distance and the Hausdorff distance.

A similarity measure $S(A,B)$, or a distance $D(T(A),T(B))$, is metric when it obeys the following:

1. $S(A,B) \geq 0$ or $D(T(A),T(B)) \geq 0$,
2. $S(A,B) = 0$ or $D(T(A),T(B)) = 0$ if, and only if, $A = B$,
3. $S(A,B) = S(B,A)$ or $D(T(A),T(B)) = D(T(B),T(A))$,
4. $S(A,C) \leq S(A,B) + S(B,C)$ or
   $D(T(A),T(C)) \leq D(T(A),T(B)) + D(T(B),T(C))$.

Many studies on image similarity [21] suggest that practical and psychologically valid measures of similarity often obey the first three conditions but do not obey the fourth condition (known as the triangle inequality) and are therefore non-metric.

In designing the similarity measure we choose the transformation T according to what image characteristics are important to the user. The following example illustrates these concerns for a stereo matching algorithm.

Example 14.2. Stereo matching Algorithm [11]. In a stereo matching, or disparity, algorithm we compare two images A and B which are two views of the same scene taken from slightly different viewing angles. The image similarity measures $S(A,B)$ should therefore be insensitive to changes due to specular reflections, occlusions, depth discontinuities and projective distortions [5, 11]. At the same time, $S(A,B)$ should be sensitive to any other changes in A and B.

It is should be clear that there is no universal similarity measure which can be used in all applications. In selecting a suitable similarity measure we find it useful to broadly divide them into two groups:

Global Measure. These measures return a single similarity value which describes the overall similarity of the two input images. The global measures may be further divided into measures which require the input images to be spatially registered and those which do not require the input images to be spatially registered.

Local Measures. These measures return a similarity image or map which describes the local similarity of the two input images. By definition the local similarity measures require the input images to be spatially registered.

Although useful, the above division of the similarity measures into two classes should not be regarded as absolute. In many cases, we may convert a global similarity measure into a local similarity measure and vice versa. The following example illustrates a case of the former.

Example 14.3. Global to Local Similarity Measures. The mean square error (mse) is a simple global similarity measure. Given two $M \times N$ spatially registered input images A and B, the global mse measure is defined as follows:

$$mse_G = \sum_{m=1}^{M} \sum_{n=1}^{N} (A(m,n) - B(m,n))^2 / MN .$$

Clearly we may apply the mse measure to individual pixels. In this case we obtain a local mse map $mse_L(m,n)$ which is defined as follows:

$$mse_L(m,n) = (A(m,n) - B(m,n))^2 .$$

More generally, we may calculate the mse over a local window. Let $W(m,n)$ define a $L \times L$ window centered at (m,n), where we assume L is an odd number (Fig. 14.1). Then we may designate the gray-levels of the pixels in A which lie in $W(m,n)$ as $\widetilde{A}(p,q|m,n), p,q \in \{1,2,\ldots,L\}$, where

$$\widetilde{A}(p,q|m,n) = A(m+p-1-\lfloor L/2 \rfloor, n+q-1-\lfloor L/2 \rfloor) .$$

Similarly, $\widetilde{B}(p,q|m,n), p,q \in \{1,2,\ldots,L\}$ designates the gray-levels of the pixels in B which lie in $W(m,n)$. In this case, we define a local windowed mse map, $mse_W(m,n)$, as follows:

$$\begin{aligned} mse_W(m,n) &= \sum_{p=1}^{L} \sum_{q=1}^{N} \left(\widetilde{A}(p,q|m,n) - \widetilde{B}(p,q|m,n)\right)^2 / L^2 , \\ &= \sum_{p=m-\lfloor L/2 \rfloor}^{m+\lfloor L/2 \rfloor} \sum_{q=n-\lfloor L/2 \rfloor}^{n+\lfloor L/2 \rfloor} (A(p,q) - B(p,q))^2 / L^2 . \end{aligned}$$

Example 14.4. Local to Global Similarity Measures. Given two spatially registered $M \times N$ binary images A and B, a local similarity algorithm returns a local similarity measure $S_L(m,n)$ for each pixel $(m,n), m \in \{1,2,\ldots,M\}, n \in \{1,2,\ldots,N\}$. We may obtain a global similarity measure by aggregating the $S_L(m,n)$ values, e. g. by finding the global maximum of the $S_L(m,n)$ values:

$$S_G = \max_{(m,n)} (S_L(m,n)) .$$

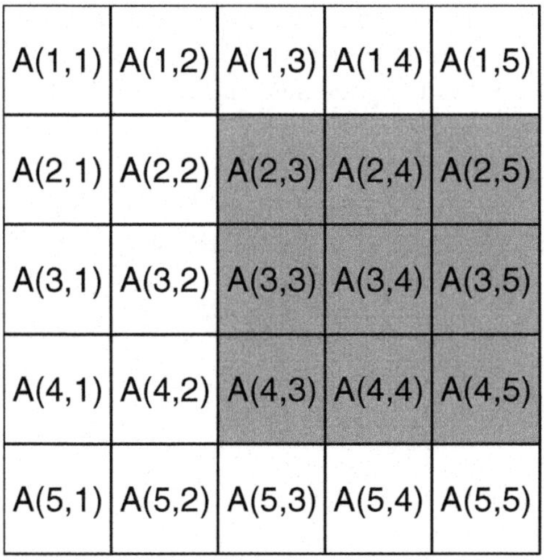

A(1,1)	A(1,2)	A(1,3)	A(1,4)	A(1,5)
A(2,1)	A(2,2)	A(2,3)	A(2,4)	A(2,5)
A(3,1)	A(3,2)	A(3,3)	A(3,4)	A(3,5)
A(4,1)	A(4,2)	A(4,3)	A(4,4)	A(4,5)
A(5,1)	A(5,2)	A(5,3)	A(5,4)	A(5,5)

Fig. 14.1 Shows an input image with gray-levels $A(m,n), m,n \in \{1,2,\ldots,5\}$. Centered at $(3,4)$ we have a 3×3 window $W(3,4)$. The gray-levels in $W(3,4)$ are $\widetilde{A}(3,3) = \{A(2,3), A(2,4), A(2,5), A(3,3), A(3,4), A(3,5), A(4,3), A(4,4), A(4,5)\}$.

We start our discussion with the global similarity measures which do not require image registration.

14.2 Global Similarity Measures without Spatial Alignment

In this section we consider global similarity measures which do not require spatial alignment. These are similarity measures which compare the probability distributions or gray-level histograms of the two images. In general these similarity measures are robust against changes in illumination. However, because they do not require spatial alignment, their discrimination power is low.

14.2.1 Probabilistic Similarity Measures

The probabilistic similarity measures are global measures which do not require the input images to be spatially registered. By converting the input images to probability distributions, they are robust against changes in illumination and are widely used when the images have been captured under widely varying illumination and viewing conditions or by different sensor types.

Let A and B denote the two input images. We convert the pixel gray-levels $a \in A$ and $b \in B$ to a common gray scale x (see Chapt. 6). Let $p(x)$ and $q(x)$ denote the probability of a transformed gray-level x appearing in A and B, then several commonly used probabilistic similarity measures are:

Chernoff

$$S_C = -\log \int_x p^{\alpha}(x) q^{1-\alpha}(x) dx , \qquad 0 < \alpha < 1 .$$

Bhattacharyya

$$S_B = -\log \int_x \sqrt{p(x) q(x)} dx .$$

Jeffrey's-Matusita

$$S_{JM} = \sqrt{\int_x \left(\sqrt{p(x)} - \sqrt{q(x)}\right)^2 dx} .$$

Kullback-Leibler

$$S_{KL} = \int_x p(x) \log \frac{p(x)}{q(x)} dx .$$

If the images A and B are spatially registered then we may use the sliding window procedure to generate local probabilistic similarity maps. Let $W(m,n)$ define a local $L \times L$ window centered at (m,n). If $\widetilde{A}$ and $\widetilde{B}$ designate the gray-levels of the pixels in A and B which lie in $W(m,n)$ (cf. Ex. 14.3) and $\widetilde{p}(x)$ and $\widetilde{q}(x)$ designate the corresponding transformed local (window) probability densities, then the local probabilistic similarity maps are

$$\begin{aligned}
\widetilde{S}_C(m,n) &= \int_x \widetilde{p}^{\alpha}(x) \widetilde{q}^{1-\alpha}(x) dx , \\
\widetilde{S}_B(m,n) &= \int_x \sqrt{\widetilde{p}(x) \widetilde{q}(x)} dx , \\
\widetilde{S}_{JM}(m,n) &= \sqrt{\int_x \left(\sqrt{\widetilde{p}(x)} - \sqrt{\widetilde{q}(x)}\right)^2 dx} , \\
\widetilde{S}_{KL}(m,n) &= \int \widetilde{p}(x) \log \frac{\widetilde{p}(x)}{\widetilde{q}(x)} dx ,
\end{aligned}$$

Note The typical window size used in the local probability similarity measures is 20×20. This is needed to ensure we have sufficient pixels to accurately calculate the local probability densities $\widetilde{p}(x)$ and $\widetilde{q}(x)$.

Example 14.5. Color Image Segmentation [17]. The goal of image segmentation is to decompose the input image into a set of meaningful or spatially coherent regions sharing similar attributes. The algorithm is often a crucial step in many video and computer vision applications such as object localization or recognition. A simple image segmentation is the K-means cluster algorithm in which we divide the pixels into K clusters. Given a input image I in a given color space, we may characterize each pixel (x,y) in I by its local histogram

$\widetilde{\mathbf{H}}(x,y)$:

$$\widetilde{\mathbf{H}}(x,y) = (\widetilde{H}_1, \widetilde{H}_2, \ldots, \widetilde{H}_M)^T ,$$

where $\widetilde{H}_m$ is the number of pixels in the local window $W(x,y)$ whose color values fall in the mth bin. We then apply the K-means algorithm as follows. Initially we define K cluster centers by randomly selecting K histograms. Let $C_1, C_2, \ldots, C_K$ denote the K cluster centers or histograms. Each pixel (x,y) is associated with a given cluster

$$\delta_k(x,y) = \begin{cases} 1 \text{ if } (x,y) \text{ is associated with } C_k , \\ 0 \text{ otherwise} . \end{cases}$$

Then the K-means algorithm attempts to find the set of cluster centers $C_k, k \in \{1,2,\ldots,K\}$, such that the overall error

$$E = \sum_{(x,y)} \sum_{k=1}^{K} \delta_k(x,y) D\left(\widetilde{H}(x,y), C_k\right) ,$$

is a minimum, where $D(\widetilde{H}(x,y), C_k)$ is an appropriate distance (similarity) measure between $\widetilde{H}(x,y)$ and C_k. Mignotte [17] recommends using the Bhattacharyya distance.

14.2.2 χ^2 Distance Measure

If we represent the transformed distributions $p(x)$ and $q(x)$ as discrete distributions p_k and q_k, then we may use the χ^2 distance as a dissimilarity measure.

Let A and B denote two input images with gray-levels $a \in A$ and $b \in B$. We convert the gray-levels $a \in A$ and $b \in B$ to a discrete common scale x by defining K pairs of corresponding bins $[a'_k, a''_k)$ and $[b'_k, b''_k)$. Then the χ^2 distance between gray-level distributions of A and B is

$$\chi^2 = \sum_{k=1}^{K} \frac{(m_k - n_k)^2}{m_k + n_k} , \tag{14.1}$$

where m_k is the number of gray-levels $a \in A$ which fall in the interval $[a'_k, a''_k)$ and n_k is the number of gray-levels $b \in B$ which fall in the interval $[b'_k, b''_k)$.

Example 14.6. Face Recognition With Local Binary Patterns [1]. Ref. [1] describes an efficient image representation based on the local binary pattern (LBP) texture features (see Sect. 3.4). Given a training set of K facial images $A^{(k)}, k \in \{1,2,\ldots,K\}$, we divide each image A_k into R regions. *Note*: We assume the training images are spatially aligned.

For each pixel (m,n) in $A^{(k)}$ we extract its local binary pattern $LBP(m,n)$ which is a label $l, l \in \{1,2,\ldots,L\}$. Then for each region $r, r \in \{1,2,\ldots,R\}$, we construct a histogram (vector)

$$\mathbf{H}_r^{(k)} = (H_r^{(k)}(1), H_r^{(k)}(2), \ldots, H_r^{(k)}(L))^T ,$$

where $H_r^{(k)}(l)$ is the number of pixels in the rth region of A_k which have a LBP label equal to l. Each training image $A^{(k)}$ is thus represented by R histograms $\mathbf{H}_r^{(k)}, r \in \{1,2,\ldots,R\}$. Given a test image B we spatially align it to the training images and carry out the above process. Let $\mathbf{h}_r, r \in \{1,2,\ldots,R\}$, denote the corresponding histograms (vectors). Then we identify the test image B as belonging to the k^*th individual if

$$k^* = \arg\min_k \left(\sum_{r=1}^{R} \chi^2(\mathbf{h}_r, \mathbf{H}_r^{(k)}) \right) ,$$

where

$$\chi^2(\mathbf{h}_r, \mathbf{H}_r^{(k)}) = \sum_{l=1}^{L} \frac{(h_r(l) - H_r^{(k)}(l))^2}{h_r(l) - H_r^{(k)}(l)} .$$

The method showed high performance on difficult face recognition experiments.

The original χ^2 distance as defined in (14.1) defined between histograms $H_A = (m_1, m_2, \ldots, m_K)$ and $H_B = (n_1, n_2, \ldots, n_K)$ and not between the discrete probability distributions p_k and q_k. However, (14.1) may be easily converted to a probability distance measure by replacing m_k and n_k by $p_k = m_k/M$ and $q_k = n_k/N$, where $M = \sum_{k=1}^{K} m_k$ and $N = \sum_{k=1}^{K} n_k$.

Example 14.7. Probability Binning [19]. In probability binning we use variable width bins such that each bin contains the same relative number of observations of A. If $p_k = m_k/M$ and $n_k = q_k/N$ denote, respectively, the relative number of observations of A and B in the kth bin, where $M = \sum_{k=1}^{K} m_k$ and $N = \sum_{k=1}^{N} n_k$, then $m_1 = m_2 = \ldots = m_K$ and the probability binning χ^2 test is

$$\chi_{PB}^2 = \sum_{k=1}^{K} \frac{|p_k - q_k|^2}{p_k + q_k} .$$

Given χ_{PB}^2 we can define a normalized scale for it as follows. Let

$$T(\chi_{PB}^2) = \max\left(0, \frac{\chi_{PB}^2 - \mu}{\sigma}\right) ,$$

then $T(\chi^2_{PB})$ represents the difference between the probability distributions of A and B as the number of standard deviations above μ, where $\mu = K/\min(M,N)$ is the minimum difference between F and G for which a confident decision of histogram difference can be made and $\sigma = \sqrt{K}/\min(M,N)$ is an appropriate standard deviation for χ^2_{PB}. Recently, Baggerly [3] has proposed a more accurate scale as follows:

$$T(\chi^2_{PB}) = \frac{\frac{2MN}{M+N}\chi^2_{PB} - (K-1)}{\sqrt{2(K-1)}} .$$

As in (14.1) we may define a local χ^2 measure:

$$\widetilde{\chi}^2(x,y) = \sum_{k=1}^{K} \frac{(\widetilde{m}_k(x,y) - \widetilde{n}_k(x,y))^2}{\widetilde{m}_k(x,y) + \widetilde{n}_k(x,y)} ,$$

where $\widetilde{m}_k(x,y), \widetilde{n}_k(x,y)$ denote, respectively, the number of pixels in $\widetilde{A}, \widetilde{B}$ which have a gray-level which falls in the kth histogram bin.

14.2.3 Cross-Bin Distance Measures

The global similarity measures considered until now (S_C, S_B, S_{JM}, S_{KL} and χ^2), all suppose the gray-levels a and b are measured on a common gray-scale. These similarity measures are therefore sensitive to any errors involved in defining the common gray-scale.

A discrete similarity measure which is less sensitive to any errors involved in defining a common gray-scale is the Earth Mover's distance (EMD) [20, 23]: Let m_k and n_k be the number of pixels in A, B which fall, respectively, in the kth histogram bin. Then the Earth Mover's distance between $\mathbf{m} = (m_1, m_2, \ldots, m_K)^T$ and $\mathbf{n} = (n_1, n_2, \ldots, n_K)^T$, is defined as

$$d_{EMD}(\mathbf{m},\mathbf{n}) = \min_{\alpha_{h,k}} \sum_{h=1}^{K}\sum_{k=1}^{K} c(h,k) ,$$

subject to

$$\alpha_{ij} \geq 0 ,$$
$$\sum_k \alpha_{hk} = m_h ,$$
$$\sum_h \alpha_{hk} = n_k ,$$

where $c(h,k)$ is an appropriate cost function. The earth mover's distance may be understood as an optimization technique which finds the minimum transportation cost. In this case, $c(h,k)$ is the cost of moving a unit mass from the hth bin to the kth bin and α_{hk} is the number of mass units carried from h to k.

Example 14.8. Mallow's Distance [12]. If we use normalized distributions $\mathbf{p} = (p_1, p_2, \ldots, p_K)^T$, where $p_k = m_k/M$ and $\mathbf{q} = (q_1, q_2, \ldots, q_K)^T$, where $Q_k = n_k/N$, then the EMD becomes the Mallow's distance [12]. If the histograms are one-dimensional and we use the following cost function $c(i,j) = |i-j|/K$, then

$$d_{mallow}(\mathbf{p},\mathbf{q}) = d_{EMD}(\mathbf{p},\mathbf{q}) = \frac{1}{K}\sum_{k=1}^{K} |P_k - Q_k| \, ,$$

where

$$P_k = \sum_{h=1}^{k} p_k \qquad \text{and} \qquad Q_k = \sum_{h=1}^{k} q_h \, .$$

The circular EMD [18] is a variant of the EMD which is used when one of the variables is circular in nature e.g. an angle.

Example 14.9. Circular Earth Mover's Distance [18]. If $\mathbf{p}$ and $\mathbf{q}$ are one-dimensional and $c(h,k) = |h-k|/K$, then the corresponding circular EMD is:

$$d_{CEMD} = \min_{h \in \{1,2,\ldots,K\}} \left(\frac{1}{K} \sum_{k=1}^{K} |\widetilde{P}_{hk} - \widetilde{Q}_{hk}| \right) ,$$

where

$$\widetilde{P}_{hk} = \begin{cases} \sum_{i=h}^{k} p_i & \text{if } k \geq h \, , \\ \sum_{i=h}^{K} p_i + \sum_{i=1}^{k} p_i & \text{if } k < h \, , \end{cases}$$

$$\widetilde{Q}_{hk} = \begin{cases} \sum_{i=h}^{k} q_i & \text{if } k \geq h \, , \\ \sum_{i=h}^{K} q_i + \sum_{i=1}^{k} q_i & \text{if } k < h \, , \end{cases}$$

An important consideration in the χ^2 and other histogram distance measures is the optimal selection of the histogram bins. This is an important issue: If the bin width is too narrow then the histogram is very noisy while if the bin width is too wide then the histogram is too smooth. In both cases, the discrimination power of the distance measure will be adversely affected. Recently [6] have described a simple semi-empirical formula for estimating the optimal number of bins in a regular histogram. We assume the pixel gray-levels are defined in the interval $[0,1]$. If there are N pixels, then the optimal number of bins is k^*:

$$k^* = \arg\max_k \left(L(k) - R(k) \right) ,$$

where

$$L(k) = \sum_{l=1}^{k} H(l) \log_2 \left(\frac{kH(l)}{N} \right) ,$$
$$R(k) = k - 1 + (\log_2 k)^{2.5} .$$

We now consider global similarity measures which require spatial alignment of the two input images A and B.

14.3 Global Similarity Measures with Spatial Alignment

In this section we consider the family of global similarity measures which require spatial alignment of the two input images. These similarity measures tend to return values change monotonically with increasing spatial misalignment. For this reason, these similarity measures are often used for spatial alignment algorithms (see Chapt. 4). We start with the mean square error (mse) and the mean absolute error (mae) which are probably the simplest measures [2].

14.3.1 Mean Square Error and Mean Absolute Error

The mean square error (mse) and the mean absolute error (mae) are defined as follows:

$$mse = \sum_k (a_k - b_k)^2 / K ,$$
$$mae = \sum_k |a_k - b_k| / K ,$$

where a_k and b_k are, respectively, the gray-levels of the kth pixel in A and B.

The mse and mae should be used when the input images have been captured with the same sensor under similar conditions, i. e. the photometric transformation between corresponding pixel gray-levels should be close to the identity transformation. Both measures are sensitive to outliers although the mae is less sensitive (more robust). In this case, we may robustify the mse and the mae by replacing the summations in the above equations by an α-trimmed summation:

---

[2] The mse and mae increase with increasing misalignment. The correlation coefficient and mutual information decrease with increasing misalignment.

$$mse_\alpha = \frac{1}{K-2\alpha} \sum_{k=\alpha+1}^{K-\alpha} d_{(k)}^2 ,$$

$$mae_\alpha = \frac{1}{K-2\alpha} \sum_{k=\alpha+1}^{K-\alpha} |d_{(k)}| .$$

where $d_{(k)} = d_l$ if $d_l = |a_l - b_l|$ is the kth largest absolute difference and α is a small number. We often set α equal to $\lfloor K/20 \rfloor$.

14.3.2 Cross-Correlation Coefficient

The cross-correlation coefficient is defined as follows:

$$\rho = \frac{\sum_k a_k b_k}{\sqrt{\sum_k a_k^2 \sum_k b_k^2}} . \tag{14.2}$$

Sometimes we use a zero-mean cross correlation coefficient. This is defined as

$$\rho_Z = \frac{\sum_k (a_k - \bar{A})(b_k - \bar{B})}{\sqrt{\sum_k (a_k - \bar{A})^2 \sum_k (b_k - \bar{B})^2}} ,$$

where $\bar{A}$ and $\bar{B}$ are, respectively, the mean gray-levels of A and B.

The cross-correlation coefficients are more robust to changes of illumination than the mse and mae. The cross-correlation coefficient should be used when the images are captured by the same sensor and any changes in illumination may be approximated with a linear transformation. Many changes in illumination are not however linear. In this case we must use mutual information and other ordinal similarity measures.

The cross-correlation coefficients may be easily made robust against outliers [2]. For example, a robust version of (14.2) is

$$S'_{CC} = \frac{\sum_k \rho_k a_k b_k}{\sqrt{\sum_k \rho_k^A a_k^2 \sum_k \rho_k^B b_k^2}} .$$

where

$$\rho_k^A = \begin{cases} a_k & \text{if } a_k < 1.345\sigma_A , \\ 1.345\sigma_A \mathrm{sgn}(a_k) & \text{otherwise} , \end{cases}$$

$$\rho_k^B = \begin{cases} b_k & \text{if } b_k < 1.345\sigma_B , \\ 1.345\sigma_B \mathrm{sgn}(b_k) & \text{otherwise} , \end{cases}$$

$$\rho_k = \sqrt{\rho_k^A \rho_k^B} ,$$

and σ_A, σ_B are the standard deviations of the a_k and b_k values.

14.3.3 Mutual Information

The mutual information [3] between two input images A and B is defined as follows:

$$MI(A,B) = \int\int p_{AB}(a,b)\log_2 \frac{p_{AB}(a,b)}{p_A(a)p_B(b)} dxdy ,$$

where $p_A(a)$ is the probability a pixel (x,y) in A has a gray-level a, $p_B(b)$ is the probability a pixel (x,y) in B has a gray-level b and $p_{AB}(a,b)$ is the probability a pixel (x,y) in A has a gray-level a and the same pixel in B has a gray-level b.

In multi-modal applications no direct relationship between the input image intensities can be assumed. In this case, similarity measures which rely on the probabilistic relation and the distribution of the intensities in the input images is used. If the input images have been captured by different sensors or by different spectral bands, then the mutual information between two images A and B is used. Further details on MI and how it is calculated is given in Sect. 4.6.

14.3.4 Ordinal Global Similarity Measures

Ordinal global similarity measures are based on order statistics. They do not use the pixel gray-levels in A and B, but use instead the ordered gray-levels. In general, these measure are insensitive to changes in illumination if the order of the gray-levels is preserved. They are often used in applications involving change detection or in applications where the images have been captured with two different sensors.

Two classical ordinal dissimilarity measures are the Spearman ρ measure and Kendall's τ measure [10]. If A,B each contain K pixels with gray-levels $a_k, b_k, k \in \{1,2,\ldots,K\}$, then these dissimilarity measures are defined, respectively, as

$$\rho = 1 - \frac{\sum_{k=1}^{K} |r_A(k) - r_B(k)|^2}{6K(K-1)} , \tag{14.3}$$

$$\tau = \sum_{k=1}^{K}\sum_{l=1}^{K} \frac{\operatorname{sgn}(a_k - a_l)\operatorname{sgn}(b_k - b_l)}{K(K-1)} , \tag{14.4}$$

where $r_A(k)$ and $r_B(k)$ denote, respectively, the rank of the kth pixel in A and B [4] and

$$\operatorname{sgn}(u) = \begin{cases} -1 & \text{if } u < 0 , \\ 0 & \text{if } u = 0 , \\ 1 & \text{if } u > 0 . \end{cases}$$

Note: The definitions given in (14.3) and (14.4) assume no ties. For corrections necessary if ties are present see e. g. [11]. Two additional ordinal dissimilarity measures are the Kemeny-Snell d_{KS} [15] and the Bhat-Nayar [5] d_{BN} distance measures.

[3] Mutual information is a similarity measure which reaches its maximum value when $A = B$

[4] The ranks $r_A(k)$ and $r_B(k)$ are defined as follows. Suppose A and B each contain K pixels with gray-levels a_k, b_k. Then $r_A(k) = l$ if a_k is the lth largest gray-level in A and $r_B(k) = l$ if b_k is the lth largest gray-level in B.

The Kemeny-Snell distance d_{KS} compares the relative ranking of each ordered pair of locations in one image with its relative ranking in the other image. Smaller values of d_{KS} indicate more agreement between the images.

Suppose A and B both contain K pixels with gray-levels $a_k, k \in \{1,2,\ldots,K\}$, and b_k. Mathematically, d_{KS} is defined as follows:

$$d_{KS}(A,B) = \sum_{k=1}^{K}\sum_{l=1}^{K} |\phi_{kl} - \psi_{kl}| ,$$

where

$$\phi_{kl} = \begin{cases} 1 \text{ if } a_k > a_l , \\ \frac{1}{2} \text{ if } a_k = a_l , \\ 0 \text{ otherwise} , \end{cases} \quad \text{and} \quad \psi_{kl} = \begin{cases} 1 \text{ if } b_k > b_l , \\ \frac{1}{2} \text{ if } b_k = b_l , \\ 0 \text{ otherwise} . \end{cases}$$

A normalized form of d_{KS} is

$$\hat{d}_{KS} = \frac{d_{KS}}{\bar{d}_{KS}} ,$$

where $\bar{d}_{KS}$ is the value of d_{KS} if the pixel gray-levels occurring in A and B were randomly distributed among the pixel locations in the two images.

Example 14.10. Kemeny-Snell Distance [15]. Given two one-dimensional image patches

$$A = (24,12,14,7,50)^T \quad \text{and} \quad B = (30,14,13,40,4)^T .$$

The corresponding ϕ_{kl} and ψ_{kl} maps are

$$\phi_{kl} = \begin{pmatrix} 0.5 & 1 & 1 & 1 & 0 \\ 0 & 0.5 & 0 & 1 & 0 \\ 0 & 1 & 0.5 & 1 & 0 \\ 0 & 0 & 0 & 0.5 & 0 \\ 1 & 1 & 1 & 1 & 0.5 \end{pmatrix} \quad \text{and} \quad \psi_{kl} = \begin{pmatrix} 0.5 & 1 & 1 & 0 & 1 \\ 0 & 0.5 & 1 & 0 & 1 \\ 0 & 0 & 0.5 & 0 & 1 \\ 1 & 1 & 1 & 0.5 & 1 \\ 0 & 0 & 0 & 0 & 0.5 \end{pmatrix} ,$$

and the Kemeny-Snell distance is

$$d_{KS} = \sum_{k=1}^{5}\sum_{l=1}^{5} |\psi_{kl} - \phi_{kl}| = 16 .$$

The Kemeny-Snell distance measure has proven efficient and useful for content-based image retrieval applications (cf. Ex. 14.1).

14.4 Local Similarity Measures

In this section we consider the family of local similarity measures. By definition, these measures require the spatial alignment of the two input images A and B. In Ex 14.3 we explained how we may convert a global similarity measure to a local similarity measure. We may use the sliding window procedure to generate local mse, mae, correlation coefficient, mutual information and d_{KS} similarity measures. In general the windows required for these similarity measures should be at least 20×20.

We now consider the Bhat-Nayer distance measure which is, by definition, a local ordinal similarity measure. It therefore does not require such a large window size: windows of 3×3 to 13×13 are common.

14.4.1 Bhat-Nayar Distance Measure

Let $W(m,n)$ denote a $L \times L$ window centered at the pixel (m,n). If $\widetilde{A}$ and $\widetilde{B}$ denote, respectively, the image pixels which lie in $W(m,n)$ in images A and B. Then the Bhat-Nayar (BN) distance measure computes the similarity of the two windows $\widetilde{A}$ and $\widetilde{B}$ by comparing the rank permutations of their pixel gray-levels as follows.

Given the two windows $\widetilde{A}$ and $\widetilde{B}$, we rewrite them as image vectors $\widetilde{\mathbf{a}} = (\widetilde{a}(1), \widetilde{a}(2), \ldots, \widetilde{a}(K))^T$ and $\widetilde{\mathbf{b}} = (\widetilde{b}(1), \widetilde{b}(2), \ldots, \widetilde{b}(K))^T$, where $K = L^2$. The corresponding rank vectors are:

$$\widetilde{\mathbf{r}}_A = (\widetilde{r}_A(1), \widetilde{r}_A(2), \ldots, \widetilde{r}_A(K))^T \quad \text{and} \quad \widetilde{\mathbf{r}}_B = (\widetilde{r}_B(1), \widetilde{r}_B(2), \ldots, \widetilde{r}_B(K))^T \ .$$

Let $k = \widetilde{r}_A^{-1}(h)$ if $h = \widetilde{r}_A(k)$. Then we may define a composite rank vector $\widetilde{\mathbf{s}}$ as:

$$\widetilde{\mathbf{s}} = (\widetilde{s}(1), \widetilde{s}(2), \ldots, \widetilde{s}(K))^T \ ,$$

where

$$\widetilde{s}(k) = \widetilde{r}_B(h) = \widetilde{r}_B\left(\widetilde{r}_A^{-1}(k)\right) \ .$$

Informally, $\widetilde{\mathbf{s}}$ is the ranking of B with respect to the ranks of A.

The BN distance measure is then

$$\lambda_{BN} = 1 - 2\frac{\max_k(d_k)}{\lfloor K/2 \rfloor} \ ,$$

where

$$d_k = k - \sum_{h=1}^{k} J(\widetilde{s}(h), k) \ ,$$

$$J(a,b) = \begin{cases} 1 \text{ if } a \leq b \ , \\ 0 \text{ otherwise} \ . \end{cases}$$

A modified BN distance measure due to Scherer, Werth and Pinz (SWP) [22] is

$$\lambda_{SWP} = 1 - \left(\frac{\max_k(d_k)}{\lfloor K/2 \rfloor} + \frac{\sum_k d_k}{\lfloor K^2/4 \rfloor} \right) .$$

The following example illustrates the calculation of the BN and the modified BN distance measures.

Example 14.11. Bhat-Nayar Distance Measure [5, 22]. Given two one-dimensional input images

$$A = (10,20,30,50,40,70,60,90,80)^T ,$$
$$B = (90,60,70,50,40,80,10,30,20)^T ,$$

the corresponding rank vectors are

$$\mathbf{r}_A = (1,2,3,5,4,7,6,9,8)^T , \qquad \mathbf{r}_B = (9,6,7,5,4,8,1,3,2)^T .$$

The composition permutation vector $\widetilde{\mathbf{s}}$ is

$$\widetilde{\mathbf{s}} = (9,6,7,4,5,1,8,2,3)^T ,$$

and the corresponding distance vector is

$$\mathbf{d} = (1,2,3,3,3,2,2,1,0)^T .$$

The BN and the modified BN distance measures are:

$$\lambda_{BN} = 1 - 2\frac{\max_k(d_k)}{\lfloor K/2 \rfloor} = 1 - 2 \times 3/4 = -0.5 ,$$
$$\lambda_{SWP} = 1 - \left(\frac{\max_k(d_k)}{\lfloor K/2 \rfloor} + \frac{\sum_k d_k}{\lfloor K^2/4 \rfloor} \right) = 1 - (3/4 + 17/20) = -0.60 .$$

The following matlab code may be used to calculate λ_{BN} and λ_{SWP}.

Example 14.12. Matlab Code for λ_{BN} and λ_{SWP}. Let A and B be two input vectors containing M gray-levels $A_k, k \in \{1,2,\ldots,K\}$, and $B_k, k \in \{1,2,\ldots,K\}$.

```
[junk, invrA] = sort(A); [junk, rA] = sort(invrA);
[junk, invrB] = sort(B); [junk, rB] = sort(invrB);
s = rB(invrA);
S = ones(K,1) * s(:)';
G = (1:K)' * ones(1,K);
```

$d = (1:K)' - sum(tril(S < G),2);$
$\lambda_{BN} = 1 - 2 * max(d)/floor(K/2);$
$\lambda_{SWP} = 1 - (\max(d)/floor(K/2) + \Sigma(d)/floor(K^2/4));$

14.4.2 Mittal-Ramesh Ordinal Measure

Although the ordinal similarity measures are robust to monotonic changes in intensity, they are not very robust to Gaussian noise. Even a small amount of Gaussian noise can completely change the rankings between pixels that are not far from each other in gray-level. Such a drawback occurs because the ordinal similarity measure do not take into account the pixel gray-levels at all. In the Mittal-Ramesh ordinal measure we take into account the pixel gray-levels. The similarity measure has a very good performance but is computationally very expensive.

14.5 Binary Image Similarity Measure

Special similarity measures are used for binary images. Given two binary images A and B we may define a local distance measure [4] as follows: Let d_A and d_B be the corresponding distance transform images [8]:

$$\begin{aligned} d_A(m,n) &= \min_{(u,v)} \widetilde{A}(u,v)\sqrt{(m-u)^2+(n-v)^2} , \\ d_B(m,n) &= \min_{(u,v)} \widetilde{B}(u,v)\sqrt{(m-u)^2+(n-v)^2} , \end{aligned}$$

where

$$\widetilde{A}(u,v) = \begin{cases} 1 & \text{if } A(u,v) = 1 , \\ \infty & \text{otherwise} , \end{cases} \qquad \text{and} \qquad \widetilde{B}(u,v) = \begin{cases} 1 & \text{if } B(u,v) = 1 , \\ \infty & \text{otherwise} . \end{cases}$$

Then, the local distance measure is defined as:

$$L(m,n) = |A(m,n) - B(m,n)| \max(d_A(m,n), d_B(m,n)) .$$

The following example illustrates the calculation of the local distance measure

Example 14.13. Local distance map. Given two binary images

$$A = \begin{pmatrix} 0\,1\,1 \\ 1\,0\,0 \\ 1\,0\,0 \end{pmatrix} \qquad \text{and} \qquad B = \begin{pmatrix} 1\,1\,0 \\ 0\,1\,1 \\ 1\,0\,1 \end{pmatrix} ,$$

the corresponding distance transforms are

$$d_A = \begin{pmatrix} 1 & 0 & 0 \\ 0 & 1 & 1 \\ 0 & 1 & 0 \end{pmatrix} \quad \text{and} \quad d_B = \begin{pmatrix} 0 & 0 & 1 \\ 1 & 1 & \sqrt{2} \\ 0 & 1 & \sqrt{5} \end{pmatrix} .$$

and the local similarity measure distance map is

$$L = \begin{pmatrix} 1 & 0 & 1 \\ 1 & 0 & \sqrt{2} \\ 0 & 0 & \sqrt{5} \end{pmatrix} . \tag{14.5}$$

We now describe how we may convert the local binary distance measure map $L(m,n)$ into global Hausdorff measures.

14.5.1 Hausdorff Metric

The Hausdorff distance [9] between two $M \times N$ binary images A and B is defined as

$$H(A,B) = \max_{(m,n)} (L(m,n)) , \tag{14.6}$$

where $L(m,n)$ is the local distance measure defined in (14.5). According to (14.6), $H(A,B)$ is the maximum distance from a point in one image to the nearest point in the second image. It is therefore very sensitive to noise and for this reason we often use robust variants of the Hausdorff distance where we replace the maximum operator in (14.6) by robust alternatives. Some examples are:

Partial Hausdorff distance

$$H_k(A,B) = L_{(k)} ,$$

where $L_{(k)} = L(m,n)$ if $L(m,n)$ is the kth largest local distance value.

Mean Hausdorff distance

$$H_{AVE}(A,B) = \sum_{(m,n)} L(m,n)/(MN) .$$

Median Hausdorff distance

$$H_{MED}(A,B) = \text{med}_{(m,n)} (L(m,n)) .$$

The following example illustrates the calculation of the original Hausdorff distance and its variants for the two binary images A and B.

Example 14.14. Hausdorff distances. Given two binary images

$$A = \begin{pmatrix} 0 & 1 & 1 \\ 1 & 0 & 0 \\ 1 & 0 & 0 \end{pmatrix} \quad \text{and} \quad B = \begin{pmatrix} 1 & 1 & 0 \\ 0 & 1 & 1 \\ 1 & 0 & 1 \end{pmatrix},$$

the corresponding local distance map $L(m,n)$ (see Ex. 14.13) is

$$L = \begin{pmatrix} 1 & 0 & 1 \\ 1 & 0 & \sqrt{2} \\ 0 & 0 & \sqrt{5} \end{pmatrix}.$$

The original, partial, mean and median Hausdorff distance measures are, respectively, $H(A,B) = \sqrt{5}$, $H_k(A,B) = \sqrt{2}$, $H_{AVE}(A,B) = (3+\sqrt{2}+\sqrt{5})/9 \approx 0.75$, and $H_{MED}(A,B) = 1$, where $H_k(A,B)$ was calculated assuming $k = 8$.

14.6 Software

END-L1. A fast matlab routine for the earth movers distance assuming an L^1 metric. Authors: Haibin Ling and Kazunori Okada [14].

COMP STATS TOOLBOX. A computational statistics toolbox. Authors Wendy Martinez and Angel Martinez [16].

14.7 Further Reading

In this chapter we have concentrated on similarity measures which are fixed in the sense that they are not learnt from training data. Recently the training of such measures has received increasing interest. A comprehensive survey of the subject is given in [24].

References

1. Ahonen, T., Hadid, A., Pietikainen, M.: Face description with local binary patterns: application to face recognition. IEEE Trans. Patt. Anal. Mach. Intell. 28, 2037–2041 (2006)
2. Arya, K.V., Gupta, P., Kalra, P.K., Mitra, P.: Image registration using robust *M*-estimators. Patt. Recogn. Lett. 28, 1957–1968 (2007)
3. Baggerly, K.A.: Probability binning and testing agreement between multivariate immunofluorescence histogram extending the chi-squared test. Cytometry 45, 141–150 (2001)
4. Baudrier, E., Millon, G., Nicolier, F., Ruan, S.: The adaptive local Hausdorff-distance map as a new dissimilarity measure. Patt. Recogn. 41, 1461–1478 (2008)

5. Bhat, D.N., Nayar, S.K.: Ordinal measures for image correspondence. IEEE Trans. Patt. Anal. Mach. Intell. 20, 415–423 (1998)
6. Birge, L., Rozenholc, Y.: How many bins should be put in a regular histogram. Euro. Series in Appl. and Indust. Math. Prob. Stat. 10, 24–45 (2006)
7. Datta, R., Joshi, D., Li, J., Wang, J.Z.: Image retrieval: ideas, influences and trends of the new age. ACM Comp. Surveys 40, 1–60 (2008)
8. Fabbri, R., Da, L., Costa, F., Torelli, J.C., Bruno, O.M.: 2D Euclidean distance transform algorithms: A comparative survey. ACM Comp. Surveys 40 (2008)
9. Huttenlocher, D.P., Klanderman, G.A., Rucklidge, W.J.: Comparing images using the Hausdorff distance. IEEE Trans. Patt. Anal. Mach. Intell. 15, 850–863 (1993)
10. Kendall, M.: Rank Correlation Methods. Griffin (1962)
11. Kumar, S.S., Chatterji, B.N.: Robust similarity measures for stereo correspondence IE(I). Journal CP 85(2), 44–51 (2004)
12. Levina, E., Bickel, P.: The earth mover's distance is the Mallow's distance: some insights from statistics. In: Proc. IEEE Int. Conf. Comp. Vis., pp. 251–256 (2001)
13. Lew, M.S., Sebe, N., Djeraba, C., Jain, R.: Content-based multimedia information retrieval: state of the art and challenges. ACM Trans. Multimedia Comp. Comm. Appl. 2, 1–19 (2006)
14. Ling, H., Okada, K.: An efficient earth mover's distance algorithm for robust histogram comparison. IEEE Trans. Patt. Anal. Mach. Intell. 29, 840–853 (2007)
15. Luo, J., Etz, S.P., Gray, R.T.: Normalized Kemeny and Snell distance: a novel metric fo quantitative evaluation of rank-order similarity of images. IEEE Trans. Patt. Anal. Mach. Intell. 24, 1147–1151 (2002)
16. Martinez, W., Martinez, A.: Computational Statistics Handbook with Matlab, 2nd edn. CRC Press, Boca Raton (2008)
17. Mignotte, M.: Segmentation by fusion of histogram-based K-means clusters in different color spaces. IEEE Trans. Image Process. 17, 780–787 (2008)
18. Rabin, J., Delon, J., Gousseau, Y.: Circular Earth Mover's Distance for the comparison of local features. In: Proc. Int. Conf. Patt. Recogn. (2008)
19. Roederer, M., Treister, A., Moore, W., Herzenberg, L.A.: Probability binning comparison: a metric for quantitating univariate distribution differences. Cytometry 45, 37–46 (2001)
20. Rubner, Y., Tomasi, C., Guibas, L.J.: The earth mover's distance as a metric for image retrieval. Int. J. Comp. Vis. 40, 99–121 (2000)
21. Santini, S., Jain, R.: Similarity Measures. IEEE Trans. Patt. Anal. Mach. Intell. 21, 871–883 (1999)
22. Scherer, S., Werth, P., Pinz, A.: The discriminatory power of ordinal measures - towards a new coefficient. In: Proc. IEEE Conf. Comp. Vis. Patt. Recogn. (1999)
23. Serratosa, F., Sanroma, G.: A fast approximation of the earth-movers distance between multidimensional histograms. Int. J. Patt. Recogn. Art. Intell. 22, 1539–1558 (2008)
24. Yang, L.: Distance metric learning: A comprehensive survey. Department of Computer Science and Engineering. Michigan State University, May 19 (2006)
25. Yu, J., Amores, J., Sebe, N., Tian, Q.: Ranking metrics and evaluation measures. Adv. Imaging Electron Phys. 144, 291–316 (2006)
26. Yu, J., Amores, J., Sebe, N., Redeva, P., Tian, Q.: Distance learning for similarity estimation. IEEE Trans. Patt. Anal. Mach. Intell. 30, 451–462 (2008)

Chapter 15
Vignetting, White Balancing and Automatic Gain Control Effects

Abstract. The subject of this chapter is a collection of miscellaneous effects which affect the brightness and color perception in a input image or in a input video. For the sake of concreteness, we shall concentrate on three important effects: Vignetting, automatic gain control and white balance. Vignetting we define as a position-dependent loss of brightness in an input image. This loss of brightness manifests itself as a gradual fading-out of an image at points near the image periphery. White balancing we define as as the adjustment of an input image to recover its true coloration. Automatic gain control effects refers to the temporal variation of the camera settings as a result of automatic gain control mechanism.

15.1 Introduction

The brightness and color perception of an input image or an input video sequence are affected by a multitude of different effects. In this chapter we shall concentrate on three effects which directly impact on many image fusion applications: vignetting, automatic gain control (AGC) effects and white balance. We define vignetting as a position-dependent loss of brightness in an input image. This loss of brightness manifests itself as a gradual fading-out of an image at points near the image periphery. We define white balancing we define as the adjustment of an input image to recover its true coloration. We define automatic gain control effects as the temporal variation of the camera settings as a result of automatic gain control mechanism.

In many applications, it is assumed that the observed image intensity value $I(x,y)$ at a point (x,y) in the image plane directly reflects the scene radiance of the corresponding point (X,Y,Z) in the three-dimensional scene. However, this assumption may not hold due to vignetting and changes in the camera exposure and AGC. The result is that there are often significant color inconsistencies between input images.

15.2 Vignetting

In general, the amount of light (radiance) hitting the image plane in an image sensor varies spatially. This causes a fade-out, or reduction in image intensity as we move towards the image periphery. The spatial variation in image intensity is known as vignetting and is due to several factors, including:

Aperture Effect. This refers to the blocking of part of the incident light rays by the aperture in the image sensor. The effect of vignetting increases as the effective size of the aperture increases and vice versa.

Cosine-Fourth Law. This refers to the relationship between the radiance (L) and the irradiance (E) [1] which is derived using a simple thin lens camera model.

$$E = \frac{L\pi R^2 \cos^4 \alpha}{4d^2} . \tag{15.1}$$

Eq. (15.1) shows that the irradiance is proportional to the radiance but it decreases as cosine-fourth of the angle α that a ray makes with the optical axis. In the equation, R is the radius of the lens and d denotes the distance between the lens and the image plane.

Pupil Aberration. This refers to the nonlinear refraction of the light rays which results in a significantly nonuniform light distribution across the aperture of the image sensor.

We model the vignetting effects as

$$E(r) = V(r)L(R) .$$

15.2.1 Vignetting Correction

In general vignetting correction begins by setting up a uniform white illumination source with a known input intensity level over a reference object with low specular reflection. The camera is pointed toward the reference surface and intensity response at each pixel position is recorded. Subsequently a correction factor at each pixel position is calculated with the following form:

$$I_{LUT}(x,y) = \max_{(u,v)} (I_{REF}(u,v)) / I_{REF}(x,y) ,$$

where $I_{REF}(x,y)$ is the gray-level intensity value at (x,y) and $I_{LUT}(x,y)$ is the corresponding correction factor which is stored in a look-up table (LUT). Thereafter

[1] *Irradiance* describes the power of the light energy that falls on a unit area of an objects surface. The corresponding photometric quantity is *illumination*. Pixel gray-levels are thus quantized estimates of the image irradiance. *Radiance* is the power of light that is emitted from a unit surface area into some spherical angle. The corresponding photometric quantity is called *brightness*.

an image captured with the same camera is vignetting corrected [10] by multiplying pixel values of the image with the corresponding correction factors stored in the LUT:

$$I_{VIG}(x,y) = I(x,y)I_{LUT}(x,y) .$$

15.3 Radiometric Response Function

Cameras can only capture a limited dynamic range of a scene. Most cameras therefore compress the dynamic range of the scene. In doing this they introduce a nonlinearity between the recorded image intensity $I(r)$ and the image irradiance $E(r)$ (Fig. 15.1).

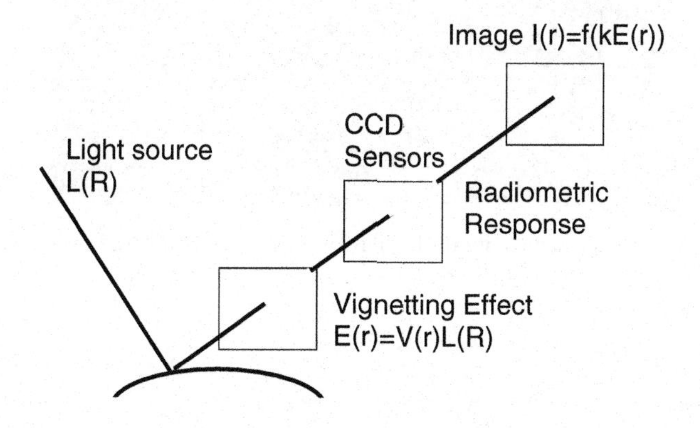

Fig. 15.1 Shows the formation of an image $I(r)$ with vignetting and radiometric response function effects

We often model the nonlinear relationship as

$$I(r) = f(kE(r)) ,$$

where k is the exposure value with which the image was captured. In general, the dynamic range of a scene exceeds that of the camera and as a consequence most cameras use an automatic gain control (AGC) to automatically adjust k so as to optimally capture the dynamic range of interest [2].

15.3.1 Automatic Gain Control

Most cameras have an automatic gain control which automatically adjusts the camera settings and camera gain k. As a consequence the same point in a three-dimensional world may appear with significantly different gray-levels in different images or video frames. Such inconsistencies are usually not negligible, even an inconsistency of 1% is noticeable in an 8-bit image as demonstrated in Fig. 15.2

Fig. 15.2 Illustrates the sensitivity of the human vision to radiometric mismatch. Consecutive image parts were biased by 3% with respect to each other. Even such a small mismatch creates clear visual artifacts [4].

Example 15.1. Rapid Gain Change in Thermal Imagery [8, 9]. Suppose we have pairs of corresponding pixels observed from two images I_1 and I_2 taken at different exposure settings (due to the AGC). We fit a parametric model to the observed correspondences to yield a "comparametric" function [7] that relates pixel intensities before and after a change in exposure. In thermal imagery we often model the radiometric response function with an affine transformation [8]:

$$f(q) = \alpha + \beta q^{\gamma} ,$$

where α, η and γ are sensor-specific constants. For an affine transformation, $\gamma = 1$. The comparametric function relating $f(q)$ to $f(kq)$ is then a straight line

$$f(k(q)) = kf(q) + \alpha(1-k) .$$

whose parameters k and α are easily found.

We describe a simple mathematical model for the AGC effect. We assume the spatial inhomogeneities introduced by the optical system have been corrected (Sect. 15.2). Then the gray-level of the pixel (x,y) is $I(x,y)$, where

$$I(x,y) = f(kL(r)) ,$$

where $L(x,y)$ is the image irradiance for the pixel (x,y), k is the exposure value with which the image was captured.

If we have two spatially aligned images of the same scene taken with different exposures k_1 and k_2, then we obtain the following relationship between the gray-levels $I_2(x,y)$ and $I_1(x,y)$:

$$I_2(x,y) = T(I_1(x,y)) ,$$

where T is known as the brightness transfer function. In many applications we assume T is a monotonically increasing function which passes through the origin. If the number of outliers is less than 5% then we may simply approximate T with a least square solution. However, when the number of outliers exceeds 5%, a robust solution is required [2].

15.4 White Balancing

The coloration on an input image often appears different depending on the illumination under which the image is taken. Different light sources have different spectral characteristics and thus require an adjustment of the captured image for the scene illuminant to recover its true coloration. White balancing aims to adjust the image automatically by finding some "white-like" areas in the image to set the parameters for balancing the colors in the remaining part of the image.

Let I be an unbalanced $M \times N$ RGB color image whose pixel components are:

$$I(x,y) = (R(x,y), G(x,y), B(x,y))^T .$$

Then white balancing aims to adjust the coloration of I to produce a balanced image I_{BAL} with pixel values $I_{BAL}(x,y) = (R_{BAL}(x,y), G_{BAL}(x,y), B_{BAL}(x,y))^T$. Since the wavelength of the G color band is close to the peak of the human luminence frequency response, we may leave the input G band unchanged, i. .e

$$G_{BAL}(x,y) = G(x,y) .$$

In the traditional gray-world white-balancing algorithm [3], we generate the $R_{BAL}(x,y)$ and $B_{BAL}(x,y)$ as follows:

$$R_{BAL}(x,y) = \bar{\alpha} R(x,y) \quad \text{and} \quad B_{BAL}(x,y) = \bar{\beta} B(x,y) .$$

The terms $\bar{\alpha}$ and $\bar{\beta}$ are global gains which are defined as follows:

$$\bar{\alpha} = \frac{\bar{G}}{\bar{R}} = \frac{\sum_k G(x_k,y_k)}{\sum_k R(x_k,y_k} \quad \text{and} \quad \bar{\beta} = \frac{\bar{G}}{\bar{B}} = \frac{\sum_k G(x_k,y_k)}{\sum_k B(x_k,y_k} ,$$

where the summation is taken over all white-like pixels in the image. Often it is difficult, or not possible, to identify the white-like pixels. In this case, we may instead use all the pixels in the input image.

In an adaptive white balancing algorithm [5, 6] we allow, $\bar{\alpha}$ and $\bar{\beta}$ to vary from pixel-to-pixel. Let $\alpha(x,y)$ and $\beta(x,y)$, be the adaptive gains, then

$$\alpha(x,y) = \frac{R(x,y)+\gamma)(\bar{G}+\gamma)^2}{G(x,y)+\gamma)(\bar{R}+\gamma)^2} \quad \text{and} \quad \beta(x,y) = \frac{B(x,y)+\gamma)(\bar{G}+\gamma)^2}{G(x,y)+\gamma)(\bar{B}+\gamma)^2} \,.$$

The corresponding corrected pixel values are:

$$\begin{aligned} R_{BAL}(x,y) &= -\gamma + \alpha(x,y)(R(x,y)+\gamma) \,, \\ G_{BAL}(x,y) &= G(x,y) \,, \\ B_{BAL}(x,y) &= -\gamma + \beta(x,y)(R(x,y)+\gamma) \,, \end{aligned}$$

where γ is a constant which is adjusted for optimum results. For a 24-bit input image, good results are obtained with $\gamma \in [50,250]$ [5, 6].

15.5 Ensemble White Balancing

The techniques of ensemble learning may be usefully employed in white balancing [1]. Instead of using one white balancing algorithm we use K different algorithms. Suppose $R_{BAL}^{(k)}(x,y), G_{BAL}^{(k)}(x,y)$ and $B_{BAL}^{(k)}(x,y)$ are the corrected colors obtained from the kth algorithm. Then, for each pixel (x,y), we fuse the $R_{BAL}^{(k)}(x,y), G_{BAL}^{(k)}(x,y), B_{BAL}^{(k)}(x,y)$ values together using an operator f:

$$\begin{aligned} \widetilde{R}_{BAL}(x,y) &= f(R_{BAL}^{(1)}(x,y), R_{BAL}^{(2)}(x,y), \ldots, R_{BAL}^{(K)}(x,y)) \,, \\ \widetilde{G}_{BAL}(x,y) &= f(G_{BAL}^{(1)}(x,y), G_{BAL}^{(2)}(x,y), \ldots, G_{BAL}^{(K)}(x,y)) \,, \\ \widetilde{B}_{BAL}(x,y) &= f(B_{BAL}^{(1)}(x,y), B_{BAL}^{(2)}(x,y), \ldots, B_{BAL}^{(K)}(x,y)) \,. \end{aligned}$$

Experiments reported in [1] show that, in general, the balanced fused values $\widetilde{R}_{BAL}(x,y)$, $\widetilde{G}_{BAL}(x,y)$ and $\widetilde{B}_{BAL}(x,y)$ represent a significant improvement over the individual $R_{BAL}^{(k)}(x,y)$, $G_{BAL}^{(k)}(x,y)$ and $B_{BAL}^{(k)}(x,y)$ values.

References

1. Bianco, S., Gasparini, F., Schettini, R.: A consensus based framework for illuminant chromaticity estimation. J. Elect. Imag. 17, 023017 (2008)
2. Kim, S.J., Pollefeys, M.: Robust radiometric calibration and vignetting correction. IEEE Trans. Patt. Anal. Mach. Intell. 30, 562–576 (2008)
3. Lam, E.Y.: Combining gray world and retinex theory for automatic white balance in digital photography. In: Proc. 9th Int. Symp. Consumer Elect. (2005)

4. Litvinov, A., Schechner, Y.Y.: Radiometric framework for image mosaicking. J. Opt. Soc. Am. 22A, 839–848 (2005)
5. Lukac, R.: Refined automatic white balancing. Elect. Lett. 43, 445–446 (2007)
6. Lukac, R.: New framework for automatic white balancing of digital camera images. Signal Process. 88, 582–593 (2008)
7. Mann, S.: Comparametric equations. IEEE Trans. Image Proc. 9(9), 1389–1406 (2000)
8. Yalcin, H., Collins, R., Hebert, M.: Background estimation under rapid gain change in thermal imagery. In: Proc. IEEE Workshop on Object tracking and classification in and beyond the visible spectrum, pp. 1142–1149 (2005)
9. Yalcin, H., Collins, R., Black, M., Hebert, M.: Background estimation under rapid gain change in thermal imagery. Comp. Vis. Image Understand. 106, 148–161 (2007)
10. Yu, W.: Practical anti-vignetting methods for digital cameras. IEEE Trans. Consumer Elect. 50, 975–983 (2004)

Chapter 16
Color Image Spaces

Abstract. The subject of this chapter is color image spaces. In the chapter we provide a brief summary of the different color spaces.

16.1 Introduction

A color space is a means by which color can be specified, created and visualized. In many applications the choice of color space is critical. The reason is that in one color space we may emphasize specific characteristics in an input image which would not be easily identified in a different color space. This is illustrated in the following example.

Example 16.1. Foreground and Shadow Detection in Traffic Monitoring [7]. Segmenting foreground objects is an important step in vehicle tracking and traffic surveillance. Ref [7] is a comparative study of different color spaces for the detection of foreground objects and their shadows in image sequences. The comparative true detection and false detection are listed in Table 16.1.

Table 16.1 Comparative True and False Detection Probabilities

Color Space	Probability of True Detection	Probability of False Detection
RGB	97.3%	0.7%
HSV	88.1%	5.8%
YC_rC_b	97.7%	0.4%
XYZ	96.7%	0.3%
rgb	91.5%	0.4%

Many different color spaces have been proposed in the literature. The commonly used color spaces may be divided into four families [11]:

Primary Systems. The primary color spaces are based on the trichronomatic theory and assume it is possible to match any color by mixing appropriate amounts of the three primary colors. Primary color spaces include RGB, XYZ and rgb. See Fig. 16.1.

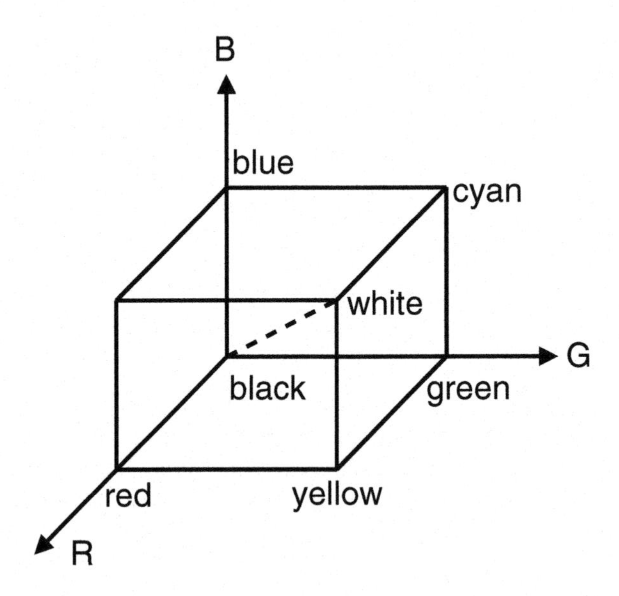

Fig. 16.1 Shows the RGB color space. Several colors are shown mapped into their location in the RGB color space.

Luminance-Chrominance Systems. The luminance-chrominance color spaces use one component to represent the luminance and two components to represent the chrominance. The luminance-chrominance spaces include YC_1C_2, AC_1C_2, $L^*u^*v^*$ and $L^*a^*b^*$.

Perceptual Systems. The perceptual spaces try to quantify the subjective human color perception by means of the intensity, hue and saturation. The perceptual color spaces include IHS, HSV, HLS and IHLS.

Statistical Independent Component Systems. The statistical independent component color spaces use statistical methods to generate components which are minimally correlated. The statistical independent component spaces include $I_1I_2I_3$ and $H_1H_2H_3$.

16.2 Perceptual Color Models

The basic process behind the perceptual color model and the transformation from an RGB coordinate system to a hue, saturation and brightness coordinate system is as follows. For 24-bit deep input image (RGB), we define the achromatic axis in RGB space as the line joining $(0,0,0)$ and $(255,255,255)$ and the chromatic plane as a plane which is perpendicular to the achromatic axis and intersects it at the origin. We then choose a function $L(C)$ which calculates the brightness or intensity of the color $C = (R,G,B)$. The projection of $L(C)$ onto the chromatic plane defines the hue and saturation of C, where the hue corresponds to the angular coordinate around the achromatic axis and the saturation corresponds to a distance from the achromatic axis. *Note*: The hue corresponds to an angular coordinate and is therefore measured in radians or degrees. Fig. 16.2 illustrates the construction of a perceptual color model.

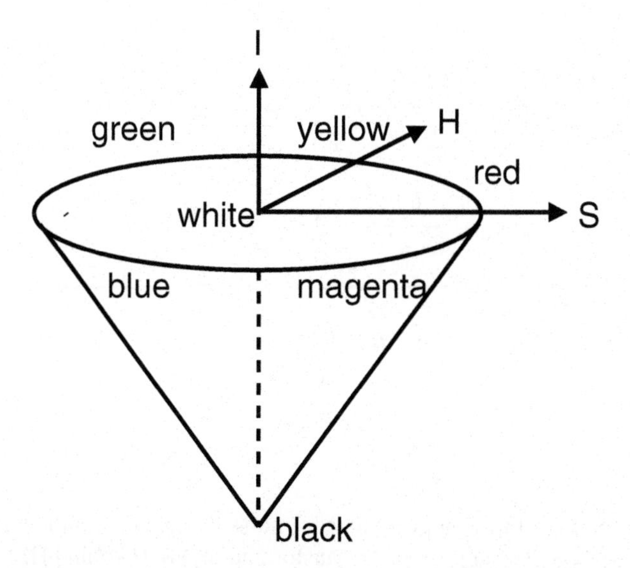

Fig. 16.2 Shows the perceptual color spaces. It has a cone shape where the central axis represents the intensity. Along this axis are all grey colors, with black at the pointed end of the cone and white at its base. The greater the distance along this axis the higher the intensity.

16.2.1 IHS

For image analysis, the most widely used perceptual color model is the IHS model. In the classical IHS model, the brightness, saturation and hue expression are:

$$L_{IHS} = \frac{1}{3}(R+G+B)\,, \tag{16.1}$$

$$S_{IHS} = 1 - \frac{3\min(R,G,B)}{R+G+B}\,, \tag{16.2}$$

$$H_{IHS} = \cos^{-1}\left(\frac{R-\frac{1}{2}(G+B)}{\sqrt{(R-G)(R-G)+(R-B)(G-B)}}\right)\,. \tag{16.3}$$

Sometimes the following algorithm is used to calculate the hue. It contains fewer multiplications and avoids the square root operation:

Example 16.2. Fast Hue Calculation.

```
if R = G = B then
    H_IHS = undefined
else
    if R ≥ B and G ≥ B then
      H_IHS = π/3 + tan^-1( √3(G − B) / (G + R − 2B) )
    else if G > R then
      H_IHS = π + tan^-1( √3(B − G) / (B + G − 2R) )
    else
      H_IHS = 5π/3 + tan^-1( √3 / (R + B − 2G) )
    end
end
```

There are also many simpler approximate formulas for calculating the IHS transformation. Two widely used approximate transformations are HSV and HLS. However [3, 4] has suggested a better approximate model is the improved HLS (IHLS) transformation.

16.2.2 HSV

The brightness function used in the HSV model is

$$L_{HSV} = \max(R,G,B)\,,$$

and the corresponding HSV saturation and hue expressions are:

$$S_{HSV} = \begin{cases} \dfrac{C_{\max} - C_{\min}}{C_{\max}} & \text{if } C_{\max} \neq 0\,, \\ 0 & \text{otherwise}\,, \end{cases}$$

$$H_{HSV} = \begin{cases} \text{undefined} & \text{if } S_{HSV} = 0\,, \\ \left(\dfrac{\pi(G-B)/3}{C_{\max} - C_{\min}} + 2\pi\right)\mathrm{mod}(2\pi) & \text{if } R = C_{\max}\,, \\ 2\pi/3 + \dfrac{\pi(B-R)/3}{C_{\max} - C_{\min}} & \text{if } G = C_{\max}\,, \\ 4\pi/3 + \dfrac{\pi(R-G)/3}{C_{\max} - C_{\min}} & \text{if } B = C_{\max}\,, \end{cases}$$

where $C_{\max} = \max(R,G,B)$ and $C_{\min} = \min(R,G,B)$.

16.2.3 HLS

The brightness function used in the HLS model is

$$L_{HLS} = \frac{C_{\max} + C_{\min}}{2}\,,$$

and the corresponding HLS model saturation and hue expressions are:

$$S_{HLS} = \begin{cases} 0 & \text{if } C_{\max} = C_{\min}\,, \\ \dfrac{C_{\max} - C_{\min}}{C_{\max} + C_{\min}} & \text{if } L_{HLS} \leq 0.5\,, \\ \dfrac{C_{\max} - C_{\min}}{2 - (C_{\max} + C_{\min})} & \text{if } L_{HLS} > 0.5\,, \end{cases}$$

$$H_{HLS} = \begin{cases} \text{undefined} & \text{if } S_{HSV} = 0\,, \\ \left(\dfrac{\pi(G-B)/3}{C_{\max} - C_{\min}} + 2\pi\right)\mathrm{mod}(2\pi) & \text{if } R = C_{\max}\,, \\ 2\pi/3\,\mathrm{deg} + \dfrac{\pi(B-R)/3}{C_{\max} - C_{\min}} & \text{if } G = C_{\max}\,, \\ 4\pi/3 + \dfrac{\pi(R-G)/3}{C_{\max} - C_{\min}} & \text{if } B = C_{\max}\,, \end{cases}$$

16.2.4 IHLS

The brightness function used in the improved HLS or IHLS model [3, 4] is

$$L_{IHLS} = 0.2126R + 0.7152G + 0.0722B ,$$

and the corresponding IHLS model saturation and hue expressions are:

$$S_{IHLS} = C_{\max} - C_{\min} ,$$
$$H_{IHLS} = \begin{cases} 2\pi - H_{HSI} & \text{if } B > G , \\ H_{HSI} & \text{otherwise} , \end{cases}$$

Apart from the above (direct) transformations, there are also indirect IHS transformations.

16.2.5 Indirect IHS Transformation

An indirect IHS transformation consists of linear transformation followed by a non-linear transformation. The following is a common indirect IHS transformation:

Linear Transformation

$$\begin{pmatrix} I \\ v_1 \\ v_2 \end{pmatrix} = \begin{pmatrix} \frac{1}{3} & \frac{1}{3} & \frac{1}{3} \\ -\frac{\sqrt{2}}{6} & -\frac{\sqrt{2}}{6} & \frac{2\sqrt{2}}{6} \\ \frac{1}{\sqrt{2}} & -\frac{1}{\sqrt{2}} & 0 \end{pmatrix} \begin{pmatrix} R \\ G \\ B \end{pmatrix} .$$

Non-linear Transformation

$$H = \tan^{-1}(v_2/v_1) ,$$
$$S = \sqrt{v_1^2 + v_2^2} .$$

The following example illustrates the merging of infrared and RGB color images using a contrast enhanced fusion method based on the linear Iv_1v_2 transformation.

Example 16.3. Merging Infrared and RGB Color Images [8]. Ref. [8] describes the contrast enhanced fusion of a color (RGB) electro-optical image EO and an infra-red image IR. The principal steps in the algorithm are as follows:

1. Transform the EO image into Iv_1v_2 space.
2. Match the infrared grayscale image IR to the electro-optical intensity image I using second-order statistics (Sect. 6.4) [8]. Let $\widetilde{IR}$ denote the transformed IR image.
3. Fuse $\widetilde{IR}$ and I using any pixel-by-pixel fusion operator (Chapt. 7). Let $\tilde{I}$ denote the fused intensity image.

4. Obtain the enhanced color image, $(\widetilde{R}\widetilde{G}\widetilde{B})$, by performing the inverse Iv_1v_2 transformation:

$$\begin{pmatrix} \widetilde{R} \\ \widetilde{G} \\ \widetilde{B} \end{pmatrix} = \begin{pmatrix} 1 & -\frac{1}{\sqrt{6}} & \frac{3}{\sqrt{6}} \\ 1 & -\frac{1}{\sqrt{6}} & -\frac{3}{\sqrt{6}} \\ 1 & \frac{2}{\sqrt{6}} & 0 \end{pmatrix} \begin{pmatrix} \widetilde{I} \\ v_1 \\ v_2 \end{pmatrix} .$$

16.2.6 Circular Statistics

In perceptual color spaces, standard statistical formula may be used to calculate statistical descriptions of the brightness and saturation values. However the hue is an angular value and so circular statistical formula must be used to calculate statistical descriptors of its values. The following example illustrates the concept of circular statistics.

Example 16.4. Circular Statistics. Given N hue values $H_i, i \in \{1,2,\ldots,N\}$, we may calculate a chrominance vector

$$C = \left(\frac{A}{N}, \frac{B}{N}\right)^T ,$$

where

$$A = \sum_{i=1}^{N} \cos H_i \qquad \text{and} \qquad B = \sum_{i=1}^{N} \sin H_i .$$

The spread of the H_i values around C is

$$V = 1 - \frac{R}{N} ,$$

where

$$R = \sqrt{A^2 + B^2} .$$

In analyzing color images we find it advantageous to use saturation weighted hue statistics. In this case, the corresponding equations are:

$$C_S = \left(\frac{A_S}{N}, \frac{B_S}{N}\right)^T ,$$

where

$$A_S = \sum_{i=1}^{N} S_i \cos H_i \qquad \text{and} \qquad B_S = \sum_{i=1}^{N} S_i \sin H_i .$$

The spread of the H_i values around C_S is

$$V_S = 1 - \frac{R_S}{N} ,$$

where

$$R_S = \sqrt{A_S^2 + B_S^2} .$$

The following example illustrates K-means clustering of the hue space.

Example 16.5. K-means clustering in Hue Space. [12]. In clustering pixels in hue space we require a distance between two hues, i. e. a distance between two angles, ϕ and θ. The simplest distance between two angles is one based on the above circular statistics formulae:

$$d_{circular}(\theta, \phi) = 1 - \frac{1}{2}\sqrt{A^2 + B^2} ,$$

where

$$A = \cos\theta + \cos\phi \qquad \text{and} \qquad B = \sin\theta + \sin\phi .$$

However, $d_{circular}(\theta, \phi)$ is non-linear and distorts the spatial relationships between the patterns. For this reason we recommend using the following linear distance [12]:

$$d(\theta, \phi) = \min(|\theta - \phi|, 2\pi - |\theta - \phi|) .$$

16.3 Multiple Color Spaces

In Ex. 16.1, we showed how the choice of a color space may emphasize specific characteristics in the input image which would not be easily identified in a different color space. We now consider the use of multiple color spaces. We start with a skin classifier (binary classification) which uses an ensemble of multiple color spaces.

Example 16.6. Skin classifier [1]. Detection of skin regions in color images is a preliminary step in many applications such as image and video classification and retrieval. Many different methods have been developed for discriminating between skin and non-skin pixels. In this example we consider the fusion of several skin classifiers which work by expressly defining the boundaries of the skin cluster in a given color space.

Among the skin classifiers are:

Hseih *et al.* [5]. Uses the IHS color space. The skin pixels satisfy at least one of the following rules: (1) $I > I_1$, $S_1 \leq S \leq S_2$ and $0 < H \leq H_1$, (2) $I > I_1$, $S_1 \leq S \leq S_2$ and $H_2 \leq H \leq 360$deg, (3) $I > I_1$, $S_3 \leq S \leq S_4$ and $H_3 \leq H \leq (H_2 - 1)$. For 24-bit color pictures [1] recommends $I_1 = 84$, $S_1 = 26$, $S_2 = 92$, $S_3 = 82$, $S_4 = 67$, $H_1 = 13$ deg, $H_2 = 337$ deg, $H_3 = 310$ deg.

Kovac *et al.* [6]. Uses the RGB space. For uniform daylight illumination the skin pixels satisfy all of the following rules: (1) $R > R_1, G > G_1, B > B_1$, (2) $\max(R,G,B) - \min(R,G,B) < \Delta$, (3) $|R-G| > L, R > G, R > B$. For flash-light illumination the rules are: (1) $R > R_2, G > G_2, B > B_2$, (2) $|R-G| \leq L$, $B < R, B < G$. For 24-bit color pictures [1] recommends $R_1 = 111, G_1 = 77$, $B_1 = 33, \Delta = 47, L = 29, R_2 = 191, G_2 = 251, B_2 = 196$.

Tsekeridou and Pitas [10]. Uses the HSV color space. The skin pixels satisfy all of the following rules: (1) $V \geq V_1$, (2) $S_1 < S < S_2$, (3) $0 \leq H \leq H_1$ or $H_2 \leq H < 360$deg. For 24-bit color pictures [1] recommends $V_1 = 52$, $S_1 = 0.25$, $S_3 = 0.64$, $H_1 = 35$ deg, $H_2 = 349$ deg.

Gomez and Morales [2]. Uses the rgb color space. This is defined as follows: $r = R/(R+G+B), g = G/(R+G+B), b = B/(R+G+B)$. The skin pixels satisfy all of the following rules: (1) $r/g > k_1$, (2) $rb/(r+g+b)^2 > k_2$ and (3) $rg/(r+g+b)^2 > k_3$. For 24-bit color pictures [1] recommends $k_1 = 1.148, k_2 = 0.054, k_3 = 0.128$.

Each of the above classifiers generates a binary map $B_k(x,y)$, where

$$B_k(x,y) = \begin{cases} 1 \text{ if } k\text{th classifier declares pixel } (x,y) \text{ a skin pixel} , \\ 0 \text{ otherwise} . \end{cases}$$

The individual pixel classifications may then be combined using the majority vote operator:

$$\widetilde{B}(x,y) = \begin{cases} 1 \text{ if } \sum_k B_k(x,y) \geq K/2 , \\ 0 \text{ otherwise} . \end{cases}$$

16.4 Software

COLOR SPACE CONVERTER. Matlab m-file for color space conversion. Available from Matlab central directory. Author: Pascal Getreuer.

16.5 Further Reading

Ref. [9] discusses the issues involved in selecting different color spaces for image feature detection.

References

1. Gasparini, F., Corchs, S., Schettini, R.: Recall or precision-oriented strategies for binary classification of skin pixels. J. Elect. Imag. 17, 023017 (2008)
2. Gomez, G., Morales, E.F.: Automatic feature construction and a simple rule induction algorithm for skin detection. In: Proc. ICML Workshop Mach. Learn. Comp. Vis., pp. 31–38 (2002)
3. Hanbury, A.: A 3D-polar coordinate color representation well adapted to image analysis. In: Proc. Scandinavian Conf. Image Analy., pp. 804–811 (2003)
4. Hanbury, A.: Constructing Cylindrical Coordinate Colour Spaces, Patt. Recogn. Lett. 29, 494500 (2008)
5. Hseih, I.-S., Fan, K.-C., Line, C.: A statistic approach to the detection of human faces in color nature scene. Patt. Recogn. 35, 1583–1596 (2002)
6. Kovac, J., Peer, P., Solina, F.: 2D versus 3D color space face detection. In: Proc. 4th EURASIP Conf. Video Image Process. Multimedia Commun., pp. 449–454 (2003)
7. Kumar, P., Sengupta, K., Lee, A., Ranganath, S.: A comparative study of different color spaces for foreground and shadow detection for traffic monitoring system. In: Proc. IEEE 5th. Int. Conf. Intell. Transport. Systems (2002)
8. Li, G., Wang, K.: Merging infrared and color visible images with a contrast enhanced fusion method. In: Proc. SPIE, vol. 6571, p. 657108 (2007)
9. Stockman, H., Gevers, T.: Selection and fusion of color models for image feature detection. IEEE Trans Patt. Anal. Mach. Intell. 29, 371–381 (2007)
10. Tsekeridou, S., Pitas, I.: Facial deature extraction in frontal views using biometric analogies. In: Proc. IX Euro. Signal Proc. Conf., vol. 1, pp. 315–318 (1998)
11. Vandenbroucke, N., Macaire, L., Postaire, J.-G.: Color image segmentation by pixel classification in an adapted hybrid color space. Application to soccer image analysis. Comp. Vis. Image Understand. 90, 190–216 (2003)
12. Vejmelka, M., Musilek, P., Palus, M., Pelikan, E.: K-means clustering for problems with periodic attributes. Int. J. Patt. Recogn. Art. Intell. 23, 721–743 (2009)

Chapter 17
Markov Random Fields

Abstract. The subject of this chapter is the *Markov Random Field* (MRF) and its use in image fusion. A Markov random field is a probabilistic model defined by local conditional probabilities. Markov random field (MRF) theory thus provides a convenient and consistent way for modeling context dependent entities such as image pixels and correlated features. Contextual models are one way to model prior information and MRF theory can be applied to model a prior probability of contextual dependent patterns. Maximum *a posteriori* (MAP) probability is one of the most popular criteria for optimality and widely applied for MRF modeling.

17.1 Markov Random Fields

A guiding insight underlying most of the work on the Markov random field (MRF) [4, 5] in image processing is that the information contained in the local physical structure of images is sufficient to obtain a good global image representation. This notion is captured by means of a local conditional probability distribution. Here the image intensity at a particular location depends only on a neighborhood of pixels. The conditional distribution is called an MRF.

The probability distributions generated by MRF's have a local neighborhood structure S. Two neighborhood systems commonly used by MRF's are depicted in Fig. 17.1. We associate an image G with a random process G whose elements are $G(m,n)$, where (m,n) refers to a pixel location in the input image. The local conditional distribution can be written as follows:

$$p\big(G(m,n)|\widetilde{G}(m,n)\big) ,$$

where $G(m,n)$ denotes the gray level of the pixel (m,n) and $\widetilde{G}(m,n)$ denotes the gray-levels of the pixels in the neighborhood S centered on the pixel (m,n) (Fig. 17.1). Although theoretically there is no restriction on the size of S, we usually choose a local neighborhood in keeping with the spirit of the MRF.

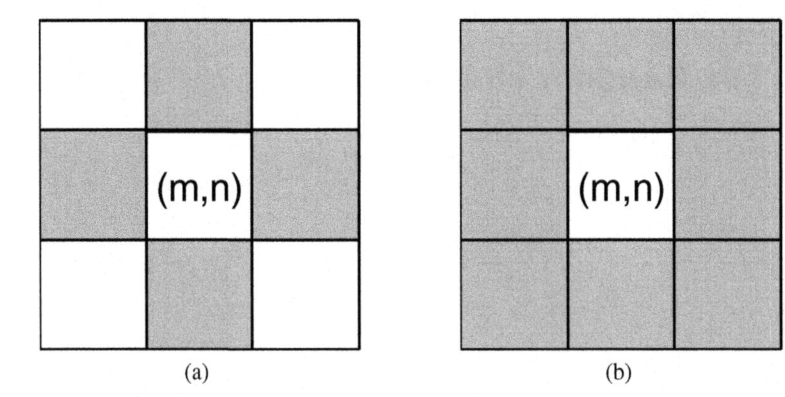

Fig. 17.1 (**a**) Shows a first-order local neighborhood structure S_1. There are four pixels in S_1 which are shown shaded. (**b**) Shows a second-order local neigborhood structure S_2. There are eight pixels in S_2 which are shown shaded.

Example 17.1. First-Order Gauss-Markov Model. Consider the pixel (m,n) in the input image G. The first-order neighborhood of (m,n) is the collection $(m,n+1)$, $(m,n-1)$, $(m+1,n)$ and $(m-1,n)$ and the corresponding conditional density takes the form

$$p(G(m,n)|\widetilde{G}(m,n)) \sim \exp - (G(m,n) - \bar{G}(m,n))^2 ,$$

where

$$\bar{G}(m,n) = (G(m,n+1) + G(m,n-1) + G(m-1,n) + G(m+1,n))/4 .$$

To make our discussion more concrete we shall concentrate on fusion of the K binary images using a Markov random field (MRF) model. The description closely follows that given in [2, 3] . Let $A_k, k \in \{1,2,\ldots,K\}$, denote a set of K binary images which have been generated by thresholding a gray-level image G with K different thresholding algorithms. Suppose B^* denotes the optimal binary image formed by fusing the A_k using the maximum *a posteriori* (MAP) decision criteria:

$$B^* = \arg\max_B \left(p(B|A_1,A_2,\ldots,A_K)\right) . \qquad (17.1)$$

Although the solution in (17.1) is optimal, it is computational intractable for any reasonable sized image. By adopting the MRF approach we may, however, significantly reduce the computational complexity by exchanging the global optimization in (17.1) to a collection of local optimizations. Mathematically, the maximization in (17.1) is equivalent to the minimization of a sum of local energy functions U_{mn}:

$$\max\left(P(B|A_1,A_2,\ldots,A_k)\right) = \min\left(\frac{1}{Z}\exp - \sum_{(m,n)} U_{mn}\right),$$

where Z is a normalizing constant and U_{mn} is a local energy function. Let $\widetilde{B}(m,n)$ and $\widetilde{A}_K(m,n)$ denote the set of labels of the pixels of the image B and the input images $A_K, k \in \{1,2,\ldots,K\}$, respectively, in a predefined neighborhood centered at (m,n). Mathematically U_{mn} is a function of the labels $B(m,n), \widetilde{B}(m,n)$ and $\widetilde{A}_k(m,n), k \in \{1,2,\ldots,K\}$:

$$U_{mn} = U(B,\widetilde{B},\widetilde{A}_1,\widetilde{A}_2,\ldots,\widetilde{A}_K|m,n)\,.$$

For simplicity, we often assume the local energy function U_{mn} can be decomposed into two contributions:

Spatial Energy $U'(B,\widetilde{B}|m,n)$**.** This is a spatial energy function which measures the spatial correlation between the label of the pixel (m,n) and the labels of its neighbors in the image B.

Inter-image Energy $U''(B,\widetilde{A}_k|m,n)$**.** This is an inter-image energy function which measures the relationship between the image B and each of the input images A_k.

If we assume the spatial and inter-image contributions are separable and additive, then we may rewrite the local energy U_{mn} as:

$$U_{mn} = \beta' U'(B,\widetilde{B}|m,n) + \sum_{k=1}^{K} \beta_k'' U''(B,\widetilde{A}_k|m,n)\,,$$

where β' and β_k' represent the relative contributions of the spatial and inter-image information to U_{mn}.

17.2 Energy Function

For a given neighborhood S, we define the spatial energy function $U'(B,\widetilde{B}|m,n)$ as the number of times that $B(p,q)$ is equal to $B(m,n)$. For a second-order neighborhood,

$$U'(B,\widetilde{B}|m,n) = -\sum_{p=m-1}^{m+1}\sum_{q=n-1}^{n+1} \delta\left(B(m,n),B(p,q)\right)\,, \tag{17.2}$$

where

$$\delta(a,b) = \begin{cases} 1 \text{ if } a = b\,, \\ 0 \text{ otherwise}\,. \end{cases}$$

In a similar manner, we define the inter-image energy function $U''(B,\widetilde{A}_k|m,n)$ as the number of times that $A_k(p,q) = B(m,n)$. For a second-order neighborhood,

$$U''(B,\widetilde{A}_k|m,n) = -\sum_{p=m-1}^{m+1}\sum_{q=n-1}^{n+1} \delta\left(B(m,n),A_k(p,q)\right)\,. \tag{17.3}$$

In (17.3) we assumed all decisions (represented by $A_k(m,n)$) are equally likely. However, in practice, our confidence that $A_k(m,n)$ is correct should be less if the gray-level $G(m,n)$ is close to the threshold t_k and should be more if $G(m,n)$ is far from the threshold. We introduce these considerations into (17.3) by allowing each contribution $\delta(B(m,n),A_k(p,q))$ to have a different weight $w_k(p,q)$:

$$U''(B,\widetilde{A}_k|m,n) = -\sum_{p=m-1}^{m+1}\sum_{q=n-1}^{n+1} w_k(p,q)\delta(B(m,n),A_k(p,q)) ,$$

where we use a simple weight function

$$w_k(p,q) = 1-\exp(-\gamma|G(p,q)-t_k|) , \tag{17.4}$$

and γ is a real positive constant controlling the steepness of the weight function.

The use of the weight function $w_k(p,q)$ as defined in (17.4) aims at controlling, during the fusion process, the effect of unreliable decisions at the pixel level that can be incurred by the thresholding algorithms. The possible misleading effects of the latter are further controlled at a global image level through the inter-image parameters $\beta_k, k \in \{1,2,\ldots,K\}$, which are computed as follows:

$$\beta_k'' = \exp(-\gamma|\bar{t}-t_k|) ,$$

where $\bar{t}$ is the average thresholding value:

$$\bar{t} = \frac{1}{K}\sum_{k=1}^{K} t_k .$$

Accordingly with this global weighting mechanism, a threshold value that is statistically incompatible with those of the ensemble.

17.3 Algorithm

The MRF solution is performed iteratively. The three stages in the MRF algorithm are:

Initialization. At initialization we find B by minimizing for each pixel (m,n) the local energy function U_{mn} without the spatial energy terms, i. e. setting $\beta' = 0$

$$U_{mn} = \sum_{k=1}^{K} \beta_k'' U''(B,\widetilde{A}_k|m,n) .$$

Iteration. Update B by minimizing for each pixel (m,n) the local energy function U_{mn} as defined in (17.1).

Stop Criterion. Repeat the iterative step until the maximum number of iterations is reached or until the number of different labels in B computed over the last two iterations becomes very small.

For 8-bit deep input pictures, Melgani [2, 3] recommends: (1) $\beta' = 1$. The results were not found to be very sensitive to this parameter. (2) $\gamma = 0.1$. This generates a confidence of 0.90 for a difference value of around 25 between the threshold value and the pixel gray-level.

17.4 Further Reading

Li [1] has written an excellent introduction to the use of MRF's in image analysis.

References

1. Li, S.Z.: Markov Random Field Modelling in Image Analysis, 3rd edn. Springer, Heidelberg (2009)
2. Melgani, F.: Robust image binarization with ensembles of thresholding algorithms. J. Elect. Imag. 15, 023010 (2006)
3. Melgani, F., Bazi, Y.: Markovian fusion approach to robust unsupervised change detection in remotely sensed imagery. IEEE Trans. Geosci. Remote Sens. Lett. 3, 457–466 (2006)
4. Serico, S.B., Moser, G.: MRF-based remote sensing image classification with automatic model parameter estimation. In: Signal and Image Processing for Remote Sensing, pp. 305–326. CRC Press, Boca Raton (2006)
5. Solberg, A.H.S., Taxt, T., Jain, A.K.: A Markov random field model for classification of multisource satellite imagery. IEEE Trans. Geosci. Remote Sens. 43, 100–113 (1996)

Chapter 18
Image Quality

Abstract. The subject of the present chapter is the objective assessment of the image quality of the output image in image fusion. A brief review of the different quality measures is given.

18.1 Introduction

Objective image quality assessment plays an important role in many image fusion applications. The two main uses of an objective image quality measure are:

1. To obtain a quantitative estimate of the quality of the fused image. A good image quality metric should be consistent, accurate, and monotonic in predicting the quality of an image. In general we prefer a quality measure which correlates with perceived quality.
2. To be used in a benchmark algorithm to compare the relative performances of different image fusion algorithms.

In general image quality measures work by exploiting the pixel difference between images, the correlation between images, and changes in the histogram. We shall start by discussing quality measures which rely on a reference image R.

18.2 Reference-Based Quality Measures

The simplest approach to assessing the quality of a fused image F is to compare it with a known reference image R. For this purpose any of the similarity measures discussed in Chapt. 14 may be used. However, of these measures, in practice only the mse, mae, correlation coefficient and mutual information measures are used:

Mean square error. The mean square error (mse) between F and R is defined as:

$$mse(F,R) = \frac{1}{MN} \sum_{m=1}^{M} \sum_{n=1}^{N} \left(F(m,n) - R(m,n)\right)^2 ,$$

where $F(m,n)$ and $R(m,n)$ are, respectively, the gray-level of the pixel (m,n) in F and R.

Mean absolute error. The mean absolute error (mae) between F and R is defined as:

$$mae(F,R) = \frac{1}{MN}\sum_{m=1}^{M}\sum_{n=1}^{N}|R(m,n)-F(m,n)|^2 .$$

Mutual Information. The mutual information (MI) between F and R is defined as:

$$MI(F,R) = \int p(R,F)\log\frac{p(R,F)}{p(R)p(F)} ,$$

where $p(R)$ and $p(F)$ are the probabilty distribution of the gray-levels in R and F, and $p(R,F)$ is the joint probability distribution of the gray-levels of R and F (see Sect. 4.6).

Correlation coefficient. The correlation coefficient ρ between F and R is defined as:

$$\rho(F,R) = \frac{\sum_{m,n} F(m,n)R(m,n)}{\sum_{m,n} F(m,n)^2 + \sum_{m,n} R(m,n)^2} . \tag{18.1}$$

Additional reference quality measures which are sometimes used include:

Wang and Bovik. The Wang and Bovik quality measure Q_{WB} [6] between F and R uses a structural similarity measure as a performance measure of the fusion algorithm. The structural similarity measure is composed of three terms: the correlation coefficient between F and R, the luminence distance between F and R and the contrast difference between F and R. It is defined as follows:

$$Q_{WB}(F,R) = \frac{\sigma_{FR}}{\sigma_F\sigma_R}\frac{2\bar{F}\bar{R}}{\bar{F}^2+\bar{R}^2}\frac{2\sigma_F\sigma_R}{\sigma_F^2\sigma_R^2} ,$$

where σ_F, $\bar{F}$ and σ_R, $\bar{R}$ are, respectively, the standard deviation and mean of F and R and σ_{FR}^2 is the covariance of F and R.

Difference of Entropy. The difference in entropy (DE) between F and R is defined as:

$$DE(F,R) = |\sum_{l=0}^{L-1} p_F(l)\log_2 p_F(l) - \sum_{l=0}^{L-1} p_R(l)\log_2 p_R(l)| , \tag{18.2}$$

18.3 Non-reference Based Quality Measures

In many image fusion applications we do not have a reference image R with which to compare the fused image F. In this case, we use a non-reference quality measure. The most common non-reference quality measures are:

Xydeas and Petrovic. The Xydeas and Petrovic (Q_{XP}) [8] quality measure uses the amount of edge information which is "transferred" from the input images I_k to the fused image F as a performance measuee of the fusion algorithm. It is defined as follows. Let $S_k(m,n)$, $\theta_k(m,n)$ and $\widetilde{S}(m,n)$, $\widetilde{\theta}(m,n)$ denote, respectively, the strength and orientation of the response obtained when a given edge operator is applied to an input image $I_k, k \in \{1,2,\ldots,K\}$, and to the fused image F. Then

$$Q_{XP}(F,I_1,I_2,\ldots,I_K) = \sum_{k=1}^{K} \sum_{(m,n)} w_k(m,n)\widetilde{Q}_{XP}(m,n) \Big/ \sum_{k=1}^{K} \sum_{(m,n)} w_k(m,n) ,$$

where

$$\begin{aligned} \widetilde{Q}_{XP}(m,n) &= f(G_k(m,n), \Phi_k(m,n)) , \\ G_k(m,n) &= \max(S_k(m,n)/\widetilde{S}(m,n), \widetilde{S}(m,n)/S_k(m,n)) , \\ \Phi_k(m,n) &= 1 - 2|\theta_k(m,n) - \widetilde{\theta}(m,n)|/\pi . \end{aligned}$$

The function f combines the local edge strength $G_k(m,n)$ and the local orientation $\Phi_k(m,n)$into a single value. The weights $w_k(m,n)$ are defined as a function of the edge strength at (m,n).

Qu, Zhang and Yan (QZY). The Qu, Zhang and Yan quality measure Q_{QZY} [5] uses the mean mutual information between the input images $I_k, k \in \{1,2,\ldots,K\}$, and F as a performance measure of the fusion algorithm. It is defined as follows:

$$Q_{QZY}(F,I_1,I_2,\ldots,I_K) = \frac{1}{K} \sum_{k=1}^{K} MI(F,I_k) ,$$

where $MI(F,I_k)$ is the mutual information between F and I_k.

Hossny, Nahavandi and Creighton. The Hossny, Nahavandi and Creighton quality measure Q_{HNC} [4] is defined as

$$Q_{HNC}(F,I_1,I_2,\ldots,I_K) = \frac{1}{K} \sum_{k=1}^{K} \frac{MI(I_k,F)}{H(I_k)+H(F)} ,$$

where $H(I)$ is the entropy of the image I.

We may also use the Wang-Bovik quality measure as a non-reference quality measure:

Wang and Bovik. The Wang and Bovik non-reference quality measure Q'_{WB} is defined as:

$$Q'_{WB}(F,I_1,I_2,\ldots,I_K) = \frac{1}{K} \sum_{k=1}^{K} Q_{WB}(F,I_k) .$$

In addition there are many variants of above quality measures. The following example illustrates one of them.

Example 18.1. The Local Yang et al. Quality Measure Q_{YZWL} [9]. Ref. [9] defines a local Wang-Bovik quality measure by creating a window $W(x,y)$ at each pixel (x,y) in the input images. Let $\widetilde{F}(x,y)$ and $\widetilde{I}_k(x,y)$ denote, respectively, the pixel gray-levels in F and I_k which lie in the window $W(x,y)$. Then we define the local Wang-Bovik quality measure between F and I_k as:

$$\widetilde{Q}_{WB}(F,I_k|x,y) \equiv Q_{WB}(\widetilde{F}(x,y),\widetilde{I}_k(x,y)) .$$

By summing $\widetilde{Q}_{WB}(F,I_k|x,y)$ we obtain an overall quality measure:

$$Q_{YZWL}(F,I_1,I_2,\ldots,I_K) = \frac{1}{K}\sum_{k=1}^{K} \lambda_k(x,y)\widetilde{Q}_{WB}(F,I_k|x,y) ,$$

where $\lambda_k(x,y)$ is an adaptive weight and $\lambda_k(x,y) \geq 0$, $\sum_{k=1}^{K} \lambda_k(x,y) = 1$.

18.4 Analysis

In choosing an appropriate quality measure it is important to consider its performance for the task in hand. For example, the quality measures used in fusing multifocus images [1] are not necessarily the same as those qualities measures used in electro-optical and infra-red fusion [3]. Nevertheless, general purpose quality measures are available which may be used if no special-purpose quality is available. Among the general-purpose quality measures are theWang-Bovik measures Q_{WB}, Q'_{WB} and Q_{YZWL}. Experiments show that in general, Q_{WB}, Q'_{WB} and Q_{YZWL}, are in reasonable agreement with quality measures suggested by human observers. Another family of general-purpose quality measures which are widely used are measures based on information theory: *MI*, Q_{QZY} and Q_{HNC} [2]. *Note*: Recent experiments [4] suggest that Q_{QZY} may be biased towards high entropy input images. To correct for this bias the recommendation is to use Q_{HNC} instead of Q_{QZY}.

18.5 Software

SSIM-INDEX. The Wang-Bovik similarity index [6]. Author: Zhou Wang.

18.6 Further Reading

A modern review of image quality measures is [7]. For detailed analytical study of the mutual information quality measures see [2].

References

1. Aslantas, V., Kurban, R.: A comparison of criterion functions for fusion of multi-focus noisy images. Opt. Commun. 282, 3231–3242 (2009)
2. Chen, Y., Xue, Z., Blum, R.S.: Theoretical analysis of an information-based quality measure for image fusion. Inf. Fusion 9, 161–175 (2008)
3. Dixon, T.D., Canga, E.F., Nikolov, S.G., Troscianko, T., Noyes, J.M., Canagarajah, C.N., Bull, D.R.: Selection of image fusion quality measures: objective, subjective and metric assessment. J. Opt. Soc. Am. 24A, B125–B135 (2007)
4. Hossny, M., Nahavandi, S., Creighton, D.: Comments on Information measure for performance of image fusion. Elect. Lett. 44, 1066–1067 (2008)
5. Qu, G., Zhang, D., Yan, P.: Information measure for performance of image fusion. Elect. Lett. 38, 313–315 (2002)
6. Wang, Z., Bovik, A.C., Sheikh, H.R., Simoncelli, E.P.: Image quality assessment: From error visibility to structural similarity. IEEE Trans. Image Process. 13, 600–612 (2004)
7. Wang, Z., Bovik, A.C.: Mean squared error: love it or leave it? - A new look at signal fidelity measures. IEEE Sig. Process. Mag. 26, 98–117 (2009)
8. Xydeas, C., Petrovic, V.: Objective pixel-level image fusion performance measure. Elect. Lett. 36, 308–309 (2000)
9. Yang, C., Zhang, J.-Q., Wang, X.-R., Liu, X.: A novel similarity based quality metric for image fusion. Inf. Fusion 9, 156–160 (2008)

Part III
Applications

Chapter 19
Pan-sharpening

Abstract. The subject of this chapter is pan-sharpening. Present-day remote sensors produce multi-spectral images with low spatial resolution and panchromatic images with high spatial resolution. Pan-sharpening is an image fusion application in which we we generate a multi-spectral image with high spatial resolution by fusing together the multi-spectral and panchromatic images.

19.1 Introduction

In remote sensing space-borne imagery there is a trade-off between spatial and spectral resolution. This is due to a combination of a set of observational constraints imposed by the aquisition system, detector specifications and satellite motion. Earth observation satellites provide multi-spectral and panchromatic data having different spatial, spectral, temporal and radiometric resolutions. Pan-sharpening [1] is a powerful solution providing a single image which can have all the complementary information from both the multi-spectral and panchromatic images.

Example 19.1. EOS. Earth observation satellites provide multi-spectral and panchromatic data having different spatial, spectral, temporal, and radiometric resolutions. The need for a single image, which can have all the complementary information from both the multi-spectral and panchromatic images, has increased. A multi-spectral image with high spatial resolution may provide feature enhancement, increased classification accuracy, and help in change detection. The designing of a sensor to provide both high spatial and spectral resolutions is limited by the tradeoff between spectral resolution, spatial resolution, and signal-to-noise ratio. Hence, there is an increased use of image processing techniques to combine the available high spectral resolution multi-spectral image and high spatial resolution panchromatic image to produce a synthetic image that has both high spatial and spectral resolutions. These image processing techniques are known as pan-sharpening.

In pan-sharpening we fuse together low-resolution multi-spectral images with a high resolution panchromatic image to give high-resolution multi-spectral images. Fig. 19.1 shows a three bands of a multispectral LANDSAT image (resolution 30 m) and the corresponding pan-sharpened bands (resolution 15 m). For easier comparison, we show the corresponding gray-scale images in Fig. 19.2. In this chapter we shall concentrate on the component-substitution techniques.

19.2 IHS Pan-sharpening

The most widely-used component-substitution pan-sharpening technique is based on the intensity-hue-saturation (IHS) transformation. This is, however, only used when the multi-spectral image has exactly three bands corresponding to the colors R, G and B. In the IHS transformation, the intensity I represents the total amount of the light in the image. An intensity image often has the appearance of a panchromatic image. This characteristic is utilized in the pan-sharpening technique. The main steps in the technique are the following:

1. Register the low-resolution multi-spectral image to the size as the high resolution panochromatic image in order to be superimposed. The re-sampling is usually performed using bi-cubic interpolation.
2. Transform the R, G and B bands of the multi-spectral image into the I, v_1 and v_2 components (see Sect. 16.2.5).
3. Modify the high resolution panchromatic image to take into account the spectral differences with respect to the multi-spectral image, the different atmospheric and illumination conditions etc. We usually do this by histogram matching the panchromatic image to the intensity component of the IHS representation (see Sect. 6.2). Let P denote the histogram-matched panchromatic image.
4. Replace the intensity component I by the panochromatic image P and perform the inverse transformation to obtain the pan-sharpened RGB image.

Mathematically, the IHS pan-sharpening technique is as follows. If (RGB) denote the interpolated high spatial-resolution multi-spectral image, then the corresponding (Iv_1v_2) image is given by

$$\begin{pmatrix} I \\ v_1 \\ v_2 \end{pmatrix} = \begin{pmatrix} \frac{1}{3} & \frac{1}{3} & \frac{1}{3} \\ -\frac{\sqrt{2}}{6} & -\frac{\sqrt{2}}{6} & \frac{2\sqrt{2}}{6} \\ \frac{1}{\sqrt{2}} & -\frac{1}{\sqrt{2}} & 0 \end{pmatrix} \begin{pmatrix} R \\ G \\ B \end{pmatrix} . \tag{19.1}$$

and the pan-sharpened image is:

$$\begin{pmatrix} \widetilde{R} \\ \widetilde{G} \\ \widetilde{B} \end{pmatrix} = \begin{pmatrix} 1 & -\frac{1}{\sqrt{2}} & -\frac{1}{\sqrt{2}} \\ 1 & -\frac{1}{\sqrt{2}} & -\frac{1}{\sqrt{2}} \\ 1 & \sqrt{2} & 0 \end{pmatrix} \begin{pmatrix} P \\ v_1 \\ v_2 \end{pmatrix} , \tag{19.2}$$

where P is the histogram-matched panchromatic image.

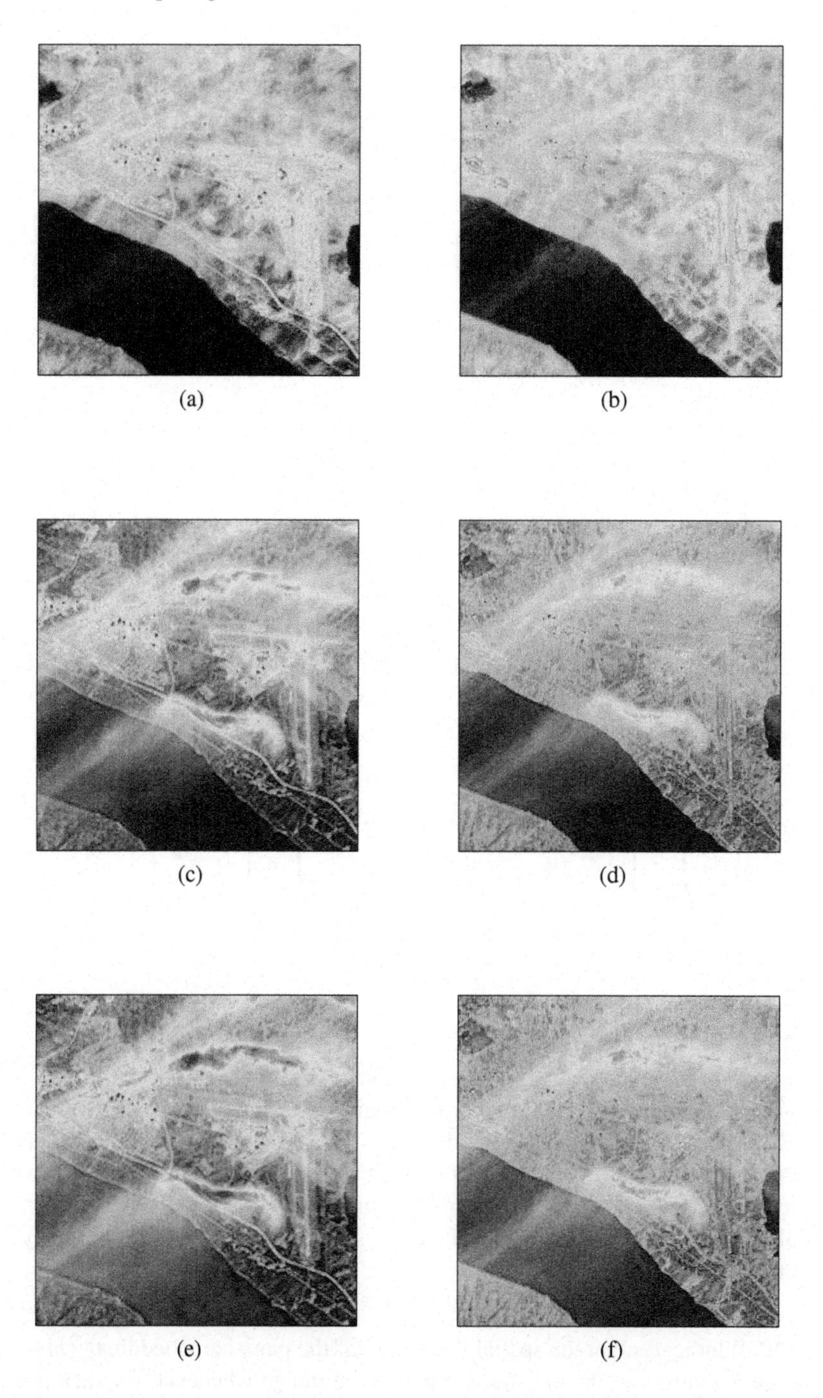

Fig. 19.1 Shows three-bands of a multi-spectral LANDSAT 7 image and the corresponding pan-sharpened bands. **(a)**, **(c)** and **(e)** Show three bands of the multi-spectral image. **(b)**, **(d)** and **(f)** Show the corresponding three bands of the pan-sharpened image. The images were created by John Childs. They are copyright of www.terrainmap.com.

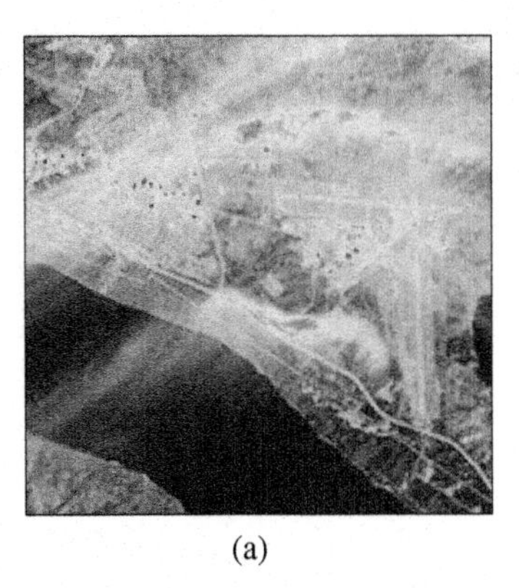

(a)

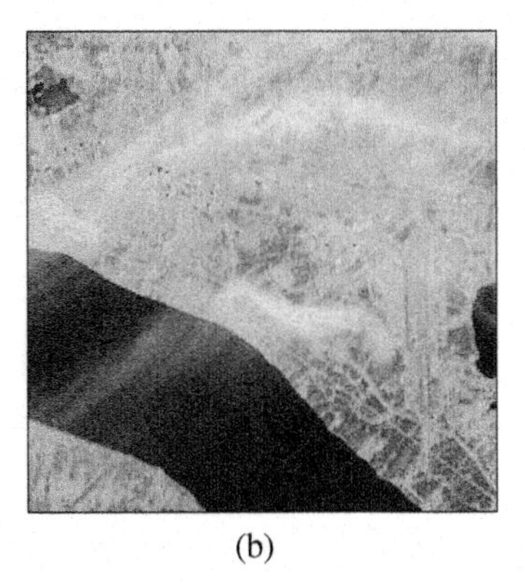

(b)

Fig. 19.2 Shows the gray-scale images corresponding to Fig. 19.1. **(a)** Shows the gray-scale image corresponding to the original multi-spectral LANDSAT images (resolution 30 m). **(b)** Shows the gray-scale image corresponding to the pan-sharpened multi-spectral image (resolution 15 m.). The original LANDSAT and pan-sharpened images were created by John Childs. They are copyright of www.terrainmap.com.

Implementing the pan-sharpening algorithm using (19.1-19.2) requires several multiplicative and additive operations. The entire process may, however, be performed with a much lower computational cost, by performing (19.1) and (19.2) in one step:

$$\begin{pmatrix} \widetilde{R} \\ \widetilde{G} \\ \widetilde{B} \end{pmatrix} = \begin{pmatrix} 1 & -\frac{1}{\sqrt{2}} & -\frac{1}{\sqrt{2}} \\ 1 & -\frac{1}{\sqrt{2}} & -\frac{1}{\sqrt{2}} \\ 1 & \sqrt{2} & 0 \end{pmatrix} \begin{pmatrix} I+(P-I) \\ v_1 \\ v_2 \end{pmatrix} = \begin{pmatrix} R+\delta \\ G+\delta \\ B+\delta \end{pmatrix} , \tag{19.3}$$

where $\delta = P - I$.

19.3 Spectral Distortion

The main drawback with the IHS pan-sharpening technique is the spectral distortion which may affect the pan-sharpened image. The primary reason for the spectral distortion appears to be the large difference between P and I. In this case, we may reduce the spectral distortion by finding a high spatial resolution image $\widetilde{I}$ which simultaneously minimizes the differences $|\widetilde{I} - P|^2$ and $|\widetilde{I} - I|^2$. Mathematically, $\widetilde{I}$ is given by

$$\widetilde{I} = \arg\min(|\widetilde{I} - P|^2 + |\widetilde{I} - I|^2) . \tag{19.4}$$

Eq. (19.4) indicates that the spatial resolution of the pan-sharpened image is higher than the resolution of the original multi-spectral image whenever, for each pixel, $\widetilde{I}$ is closer to P. At the same time, the spectral distortion of the sharpened image is higher because the difference between $\widetilde{I}$ and P is larger. By the same reasoning, the term $|\widetilde{I} - I|$ indicates that the spectral distortion of the pan-sharpened image fusion is lower whenever, for each pixel, $\widetilde{I}$ is closer to I. At the same time, the spatial resolution of the sharpened image is lower than that of the sharpened image obtained with the traditional IHS method. We see, that depending on $\widetilde{I}$, we may establish a trade-off between the spatial and spectral resolution of the pan-sharpened image. This is the basis of the pan-sharpening algorithm developed by Choi [2].

19.3.1 Pan-sharpening Algorithm of Choi

Eq. (19.4) forms the basis of the pan-sharpening algorithm of Choi [2]. We consider two special cases:

$\widetilde{I}=P$. In this case the absolute difference $|\widetilde{I}-I|$ is maximal. As a consequence, the spectral distortion is maximal and the spatial resolution is high.

$\widetilde{I}=I$. In this case the absolute difference $|\widetilde{I}-I|$ is zero. As a consequence, the spectral distortion is minimal (zero) and the spatial resolution is low.

Choi uses a parametric solution $\widetilde{I}(t)$ which is linear between the two limiting cases:

$$\widetilde{I}=P-\frac{P-I}{t}\,, \tag{19.5}$$

where t is a parameter and the limiting cases are recovered when $t=\infty$ and $t=1$. Eqn. (19.5) implies that as t increases from 1 to ∞, both the spatial resolution and the spectral distortion of $\widetilde{I}$ will increase. In the Choi algorithm, we balance the spatial resolution with the spectral resolution using the parameter t, which is called the *trade-off parameter*. The pan-sharpened image of Choi is:

$$\begin{pmatrix}\widetilde{R}\\ \widetilde{G}\\ \widetilde{B}\end{pmatrix}=\begin{pmatrix}R+\delta\\ G+\delta\\ B+\delta\end{pmatrix}=\begin{pmatrix}R+(P-I)(1-1/t)\\ G+(P-I)(1-1/t)\\ B+(P-I)(1-1/t)\end{pmatrix}, \tag{19.6}$$

where $\delta=\widetilde{I}-I=(P-I)(1-1/t)$.

The Choi method is fast and easy to implement. Moreover, according to the purpose of each application, different user-specified tradeoff parameters can be used for user-specified multi-spectral images.

To evaluate the spectral distortion in (19.6) we compare $(\widetilde{R}\widetilde{G}\widetilde{B})$ with the multi-spectral image (RGB) after conversion into the IHS color space using (16.1-16.3):

$$\widetilde{I}=\frac{1}{3}(\widetilde{R}+\widetilde{G}+\widetilde{B})=I+\delta\,,$$

$$\widetilde{H}=\cos^{-1}\left(\frac{\widetilde{R}-(\widetilde{G}+\widetilde{B})/2}{\sqrt{(\widetilde{R}-\widetilde{G})^2+(\widetilde{R}-\widetilde{B})(\widetilde{G}-\widetilde{B})}}\right)=H\,,$$

$$\widetilde{S}=1-\frac{3\min(\widetilde{R},\widetilde{G},\widetilde{B})}{\widetilde{R}+\widetilde{G}+\widetilde{B}}=\frac{IS}{\widetilde{I}}\,.$$

These equations demonstrate that while the hue of the pan-sharpened image is unchanged, the saturation is changed such that the product of the saturation and the intensity remains constant.

19.3.2 Pan-sharpening Algorithm of Tu et al.

Tu *et al.* [6] modified the pan-sharpening algorithm of Choi by multiplying the right-hand side of (19.6) by $P/\widetilde{I}$. The pan-sharpening algorithm of Tu *et al.* is thus identical to Choi apart from a simple energy normalization procedure. The corresponding Tu pan-sharpened image is:

$$\begin{aligned}\begin{pmatrix}\widetilde{R}\\ \widetilde{G}\\ \widetilde{B}\end{pmatrix} &= \frac{P}{\widetilde{I}}\begin{pmatrix}R+\delta\\ G+\delta\\ B+\delta\end{pmatrix} = \frac{P}{\widetilde{I}}\begin{pmatrix}R+(\widetilde{I}-I)\\ G+(\widetilde{I}-I)\\ B+(\widetilde{I}-I)\end{pmatrix},\\ &= \frac{P}{P(1-1/t)+I/t}\begin{pmatrix}R+(P-I)(1-1/t)\\ G+(P-I)(1-1/t)\\ B+(P-I)(1-1/t)\end{pmatrix}.\end{aligned} \tag{19.7}$$

To assess the spectral distortion of (19.7) we follow the same approach as before and obtain:

$$\begin{aligned}\widetilde{I} &= \frac{P}{\widetilde{I}}(I+\widetilde{I}-I) = P,\\ \widetilde{H} &= H,\\ \widetilde{S} &= S.\end{aligned}$$

These equations demonstrate that the hue and saturation of the pan-sharpened image of Tue *et al.* are unchanged.

19.4 IKONOS

When IHS pan-sharpened methods are used with IKONOS imagery, there is a significant color distortion, due primarily to the range of wavelengths in an IKONOS panochromatic image. Unlike the panochromatic images of SPOT and IRS sensors, IKONOS panochromatic images (as shown Fig. 2.5) have an extensive range of wavelengths-from visible to near-infrared (NIR). This difference obviously induces the color distortion problem in IHS fusion as a result of the mismatches; that is, the P and I are spectrally dissimilar. In particular, the grey values of P in the green vegetated regions are far larger than the grey values of I because the areas covered by vegetation are characterized by a relatively high reflectance of NIR and panochromatic bands as well as a low reflectance in the R, G and B bands. To minimize the radiance differences between I and panchromatic P, we may include the NIR band in the definition of the I component. A simple method for doing this is:

$$I = \frac{R + \alpha G + \beta B + NIR}{3}, \tag{19.8}$$

where $0 \leq \alpha \leq 1$ and $0 \leq \beta \leq 1$ are two parameters and $\alpha + \beta = 1$.

IHS pan-sharpening is then extended from three color image (RGB) to the four band image $(RGB) + NIR$:

$$\begin{pmatrix} \widetilde{R} \\ \widetilde{G} \\ \widetilde{B} \\ \widetilde{NIR} \end{pmatrix} = \begin{pmatrix} R + \delta \\ G + \delta \\ B + \delta \\ NIR + \delta \end{pmatrix}, \tag{19.9}$$

where $\delta = P - I$.

Without loss of generality the pan-sharpening algorithm of Choi and Tu *et al.* may be directly extended to (19.9). For example, the corresponding Tu *et al.* equations are:

$$\begin{pmatrix} \widetilde{R} \\ \widetilde{G} \\ \widetilde{B} \\ \widetilde{NIR} \end{pmatrix} = \frac{P}{P(1 - 1/t) + I/t} \begin{pmatrix} R + (P - I)(1 - 1/t) \\ G + (P - I)(1 - 1/t) \\ B + (P - I)(1 - 1/t) \\ NIR + (P - I)(1 - 1/t) \end{pmatrix}.$$

19.5 Wavelets

The basis of the IHS pan-sharpening technique is that the IHS transformation separates the spectral and spatial information of the multi-spectral RGB image. In practice the separation of the spectral and spatial information is not complete: The H and S components include some spatial information and the I component includes some spectral information. As a result IHS pan-sharpened images often differ significantly from what the multi-spectral image would have if it had been collected at the resolution of the panchromatic image. One way to reduce the distortion is to retain the I component while adding spatial details of the panchromatic image which are missing from the multi-spectral image. This is the central idea of the IHS-wavelet algorithm.

The detail information of the PAN image that corresponds to structures of features with a size between the spatial resolution of the PAN image and that of the MS one is extracted using the multi-resolution wavelet decomposition in an undecimated way.

Multi-resolution analysis allows us to decompose images into a sequence of wavelet detail planes which correspond to structures of a given size.

The idea of wavelet-IHS pan-sharpening is instead of replacing the I component with the panchromatic image we simply add spatial details of the panchromatic image which are missing from the I component. The wavelet decomposition provides us with a simple way of doing this.

The main steps in the method (assuming a spatial resolution ratio of 1 : 4) [4, 5] are as follows (see also Ex. 8.6):

1. Spatially align the multi-spectral image and the panchromatic image. Then re-sample the multi-spectral image to make its pixel size equal to that of the panchromatic image.
2. Apply the IHS transform to the multi-spectral image and obtain the corresponding intensity, hue and saturation components I, H and S.
3. Generate a new panchromatic image P whose histogram matches that of the I image.
4. Decompose P using the undecimated wavelet decomposition. We obtain two wavelet detail images $y_1^{(P)}$ and $y_2^{(P)}$ and an approximation (residual) image P_2. Repeat for I. We obtain two wavelet detail images $y_1^{(I)}$ and $y_2^{(I)}$ and an approximation (residual) image I_2. By definition, the detail images $y_1^{(P)}$ and $y_2^{(P)}$ pick up structural details which are not present I.
5. Calculate the difference $\delta = \widetilde{I} - I = \sum_k y_k^{(P)} - \sum_k y_k^{(I)}$, where $\sum_k y_k^{(P)} = y_1^{(P)} + y_2^{(P)}$ and $\sum_k y_k^{(I)} = y_1^{(I)} + y_2^{(I)}$.
6. Insert the spatial information of the panchromatic image into the multi-spectral image through the inverse IHS transform.

Mathematically,

$$\begin{pmatrix} \widetilde{R} \\ \widetilde{G} \\ \widetilde{B} \end{pmatrix} = \begin{pmatrix} R + (\widetilde{I} - I) \\ G + (\widetilde{I} - I) \\ B + (\widetilde{I} - I) \end{pmatrix} , \\ = \begin{pmatrix} R + (\sum y_k^{(P)} - \sum y_k^{(I)}) \\ G + (\sum y_k^{(P)} - \sum y_k^{(I)}) \\ B + (\sum y_k^{(P)} - \sum y_k^{(I)}) \end{pmatrix} .$$

The new saturation value for the image fused by the method of Gonzlez-Audcana *et al.* then becomes

$$\widetilde{S} = 1 - \frac{3 \min(R + \delta, G + \delta, B + \delta)}{R + G + B + 3\delta} , \tag{19.10}$$

where $\delta = \widetilde{I} - I$ and

$$\frac{\widetilde{S}}{S} = \frac{1}{\widetilde{I}} = \frac{1}{1 + \delta} . \tag{19.11}$$

19.6 Sensor Spectral Response

If we calibrate the panchromatic image P and the three color images R, G, B in the multi-spectral image, then we may introduce the spectral response $\phi(\nu)$ of the sensors into (19.3) as follows [3, 7]: We calibrate P, R, G and B in terms of the corresponding number of photons. Thus

$$P = n_P\,, \qquad R = n_R\,, \qquad G = n_G \qquad \text{and} \qquad B = n_B\,.$$

In this case, the number of photons corresponding to the intensity component I is n_I:

$$n_I = (n_R + n_G + n_B)/3\,.$$

Let $\hat{I} = \hat{n}_I$ denote the estimated intensity of the multi-spectral image if it works at the spatial resolution of the panchromatic sensor. Then the proposed pan-sharpened image [3] is:

$$\begin{pmatrix} \tilde{n}_R \\ \tilde{n}_G \\ \tilde{n}_B \end{pmatrix} = \begin{pmatrix} n_R + \delta n_R/\hat{n}_I \\ n_G + \delta n_G/\hat{n}_I \\ n_B + \delta n_B/\hat{n}_I \end{pmatrix}\,.$$

where $\delta = \hat{n}_I - n_I$ and according to (2.2):

$$\begin{aligned} \hat{n}_I &= \frac{\hat{n}_R + \hat{n}_G + \hat{n}_B}{3}\,, \\ &= \frac{n_P}{3}\left(\int \phi_R(\nu)d\nu + \int \phi_G(\nu)d\nu + \int \phi_B(\nu)d\nu\right) \Big/ \int \phi_P(\nu)d\nu\,. \end{aligned}$$

References

1. Alparone, L., Wald, L., Chanussot, J., Thomas, C., Gamba, P., Bruce, L.M.: Comparison of pansharpening algorithms: outcome of the 2006 GRS-S data fusion contest. IEEE Trans. Geosci. Remote Sensing 45, 3012–3021 (2007)
2. Choi, M.: A new intensity-hue-saturation approach to image fusion with a tradeoff parameter. IEEE Trans. Geosci. Remote Sens. 44, 1672–1682 (2006)
3. Gonzalez-Audicana, M., Otazu, X., Fors, O., Alvarez-Mozos, J.: A low computational-cost method to fuse IKONOS images using the spectral response function of its sensors. IEEE Trans. Geosci. Remote Sens. 44, 1683–1691 (2006)
4. Gonzalez-Audicana, M., Otazu, X., Fors, O., Seco, A.: Comparison between Mallat's and the 'a trous' discrete wavelet transform based algorithms for the fusion of multispectral and panchromatic images. Int. J. Remote Sens. 26, 595–614 (2005)
5. Gonzalez-Audicana, M., Saleta, J.L., Catalan, R.G., Garcia, R.: Fusion of multispectral and panchromatic images using improved IHS and PCA mergers based on wavelet decomposition. IEEE Trans. Geosci. Remote Sens. 42, 1291–1299 (2004)
6. Tu, T.-M., Cheng, W.-C., Chang, C.-P., Huang, P.S., Chang, J.-C.: Best tradeoff for high-resolution image fusion to preserve spatial details and minimize color distortion. IEEE Geosci. Remote Sens. Lett. 4, 302–306 (2007)
7. Otazu, X., Gonzalez-Audicana, M., Fors, O., Nunez, J.: Introduction of sensor spectral response into image fusion methods. Application to wavelet-basedmethods. IEEE Trans. Geosci. Remote Sens. 43, 2376–2385 (2005)

Chapter 20
Ensemble Color Image Segmentation

Abstract. The subject of this chapter is ensemble color image segmentation. This is an image fusion application in which combine several simple image segmentation algorithms to obtain a state-of-the-art image segmentation algorithm. The goal of image segmentation is to decompose the input image into a set of meaningful or spatially coherent regions sharing similar attributes. The algorithm is often a crucial step in many video and computer vision applications such as object localization or recognition. A simple image segmentation is the K-means cluster algorithm in which we divide the pixels into K clusters.

20.1 Introduction

The goal of image segmentation is to decompose the input image into a set of meaningful or spatially coherent regions sharing similar attributes. The algorithm is often a crucial step in many video and computer vision applications such as object localization or recognition.

Present-day state-of-the-art segmentation algorithms obtain a high performance by becoming more and more complex. In this chapter we describe a different approach in which a high quality segmentation is obtained by fusing together a large ensemble of simple image segmentation algorithms [1].

The new algorithm works as follows:

Ensemble. Given an input image I^* we create an ensemble of K images $I_k, k \in \{1,2,\ldots,K\}$, by transforming I^* into K different color spaces.

K-means Segmentation. Each image I_k is segmented into a L label decision image D_k using the K-means cluster algorithm, although any simple segmentation algorithm may be used instead.

K-means Fusion. Form a high quality segmented image $\widetilde{D}$ by fusing together the $D_k, k \in \{1,2,\ldots,K\}$:

$$\widetilde{D} = f(D_1, D_2, \ldots, D_K) ,$$

where f is an appropriate fusion operator. In [1] the K-means cluster algorithm is also used as the fusion operator.

We now consider in more detail each of these steps.

20.2 Image Ensemble

Given an input image I^* in RGB space we create a ensemble of K images $I_k, k \in \{1,2,\ldots,K\}$, by transforming I^* into K different color spaces S_k. Each color space emphasizes a different property. The color spaces used by Mignotte are:

RGB. This tri-chromatic space is non-linear with regard to visual perception.
HSV. This space decouples the chromatic information from shading effects.
YIQ. This space takes into account the human color charcateristics.
XYZ. This is a pscho-visually linear space.
LAB. This space approximates the human vision system. Its L component closely matches the human perception of lightness.
LUV. This a Euclidean space which yields a perceptually uniform spacing of color.

20.3 K-Means Segmentation

We segment each image I_k using the K-means cluster algorithm (see Ex. 6.8). In the conventional K-means we assign each pixel $m, m \in \{1,2,\ldots,M\}$, to a given cluster $C_l^{(k)}$:

$$\delta_l^{(k)}(m) = \begin{cases} 1 \text{ if } c_m^{(k)} \text{ is assigned to } C_l^{(k)} , \\ 0 \text{ otherwise} . \end{cases}$$

where $c_m^{(k)}$ is the representation of the mth pixel in I_k, i. e. in the color space S_k. The assignments $\delta_l^{(k)}(m)$ define the corresponding segmented image D_k:

$$D_k(m) = l \qquad \text{if} \qquad \delta_l^{(k)}(m) = 1 .$$

The K-means algorithm attempts to find the set of cluster centers C_l such that the overall error

$$E^{(k)} = \sum_m \sum_{l=1}^{L} \delta_l^{(k)}(m) d(c_m, C_l^{(k)})$$

is a minimum, where $d(c_m^{(k)}, C_l^{(k)})$ is an appropriate distance (similarity) measure between $c_m^{(k)}$ and $C_l^{(k)}$.

The cluster centers are found in an iterative procedure. In each iteration we calculate the assignment $\delta_l^{(k)}(m)$ using the cluster centers $C_l^{(k)}$ calculated in the previous iteration. The cluster centers are then recalculated using the new assignment:

```
for m = 1:M
    δ_h^(k)(m) = { 1 if d(c_m^(k), C_h^(k)) = min_l d(c_m^(k), C_l^(k))
                 { 0 otherwise;
end
for l = 1:L
    c_m^(k) = Σ_{m=1}^M δ_l^(k)(m) c_m^(k) / Σ_{m=1}^M δ_l^(k)(m);
end
```

For an image $I^{(k)}$ in RGB space, the color components (r_m, g_m, b_m) for each pixel $m, m \in \{1,2,\ldots,M\}$, are all measured on the same linear scale. In this case, an appropriate similarity (distance) measure is the Euclidean distance:

$$d(c_m^{(k)}, C_l^{(k)}) = \sqrt{(r_m^{(k)} - R_l^{(k)})^2 + (g_m^{(k)} - G_l^{(k)})^2 + (b_m^{(k)} - B_l^{(k)})^2}\ .$$

However, the Euclidean distance is *not* appropriate for the other color spaces, HSV, YIQ, XYZ, LAB and LUV, where the color components (h_m, s_m, v_m), (y_m, i_m, q_m), (x_m, y_m, z_m), (l_m, a_m, b_m) and (l_m, u_m, v_m) are not measured on the same linear scale.

At first sight it may be though very difficult to find a universal similarity (distance) measure. However, Mignotte [1] shows we may define such a similarity measure by converting the color components of each pixel into a local probability distribution.

Let $p_m^{(k)}$ be the local distribution of the pixel values in a window centered on the mth pixel in $I^{(k)}$. If $\bar{P}_l^{(k)}$ denotes the average distribution of pixel values which are assigned to the l cluster,

$$\bar{P}_l^{(k)} = \sum_{m=1}^{M} \delta_l^{(k)}(m) p_m^{(k)} / \sum_{m=1}^{M} \delta_l^{(k)}(m)\ ,$$

then Mignotte defines $d(c_m^{(k)}, C_l^{(k)})$ as

$$d(c_m^{(k)}, C_l^{(k)}) = S(p_m^{(k)}, \bar{P}_l^{(k)})\ ,$$

where S is the Bhattacharyya similarity (distance) measure.

Given the Mignotte distance measure $d(c_m^{(k)}, C_l^{(k)})$, we may cluster the pixels using the K-means cluster algorithm.

20.4 *K*-Means Fusion Operator

Mignotte also uses the K-means cluster algorithm as a fusion operator as follows. Given the segmented images $D^{(k)}, k \in \{1,2,\ldots,K\}$, let $q_m^{(k)}$ denote the local distribution of segmentation labels in a window centered on the mth pixel in $D^{(k)}$. We now concatenate the distributions $q_m^{(k)}$:

$$\widetilde{q}_m \equiv (q_m^{(1)}, q_m^{(2)}, \ldots, q_m^{(K)}) \ .$$

The concatenated distributions $\widetilde{q}_m$ may then be clustered into L_Q clusters using the K-means algorithm in the same way we clustered the $p_m^{(k)}$. The result is an assignment matrix:

$$\widetilde{\delta}_l(m) = \begin{cases} 1 \text{ if } \widetilde{q}_m \text{ is assigned to } \widetilde{C}_l \ , \\ 0 \text{ otherwise} \ . \end{cases}$$

The assignment matrix $\widetilde{\delta}_l(m)$ then defines the fused segmentation image $\widetilde{D}$:

$$\widetilde{D}(m) = l \quad \text{if} \quad \widetilde{\delta}_l(m) = 1 \ .$$

Reference

1. Mignotte, M.: Segmentation by fusion of histogram-based K-means clusters in different color spaces. IEEE Trans. Image Process. 17, 780–787 (2008)

Chapter 21
STAPLE: Simultaneous Truth and Performance Level Estimation

Abstract. The subject of this chapter is the STAPLE (Simultaneous Truth and Performance Level Estimation) algorithm. This is a method for fusing together several segmented images and is based on the expectation-maximization (EM) algorithm.

21.1 Introduction

Segmentation by non-rigid registration to an atlas image is an established method for labeling biomedical images [1]. However, in general, the choice of atlas image is important in determining the quality of the segmentation. As a result, multiple atlases are often used which may improve the segmentation accuracy over approaches that use a single individual atlas or even an average atlas (see Sect. 10.8). These algorithms often combine the multiple segmentations by a majority-vote rule. However, [2, 3] have shown that much better results may be produced by using the STAPLE (Simultaneous Truth and Performance Level Estimation) procedure . The STAPLE procedure uses the expectation-maximization (EM) algorithm to iteratively estimate the quality of the individual segmentations. The final segmentation is then computed with these individual segmentation qualities taken into account by weighting the decisions made by a reliable segmentation algorithm higher than ones made by a less reliable algorithm.

21.2 Expectation-Maximization Algorithm

The expectation-maximization (EM) is a powerful iterative technique suited for calculating the maximum likelihood estimate in problems where the observation can be viewed as incomplete data. The maximum-likelihood estimate of a random variable x, denoted as $\hat{x}$, based on the incomplete observed data y, is defined as

$$\hat{x} = \arg\max_{x}(\log p(y|x)) ,$$

where $\log p(y|x)$ is the log likelihood of y given x.

Central to the EM algorithm is the introduction of an auxiliary random variable W with log likelihood $\log p(W|x)$. The data W is referred to as the complete data because it is more informative than y. The complete data W is not observed directly, but indirectly through y via the relationship $y = f(W)$, where f is a many-to-one mapping function. Those unobserved variables are referred to as hidden data and denoted by H. The EM algorithm calculates an estimate for x, $\hat{x}$, through an iterative procedure in which the next iteration's estimate of x is chosen to maximize the expectation of $\log p(W|x)$ given the incomplete data y and the current iteration's estimate of x. For a review of the EM algorithm when we use a mixture of Gaussians for the likelihood see Sect. 7.3.

21.3 STAPLE

Given a medical input image I, our task is to estimate its ground truth segmentation vector T. Writing I as a one-dimensional vector with M pixels $I(m), m \in \{1,2,\ldots,M\}$, then

$$T(m) = \begin{cases} 1 \text{ if a given structure exists at } m\text{th pixel} , \\ 0 \text{ otherwise} . \end{cases}$$

Suppose we have segmented I using K different experts and/or algorithms. Let $B_k, k \in \{1,2,\ldots,K\}$, denote the corresponding segmentation vectors. Then our aim is to create a vector $\widetilde{B}$ by combining the individual B_k such that $\widetilde{B}$ is a good approximation to the (unknown) ground truth segmentation vector T.

We have already considered this problem in Chapts. 7 and 12. Here we show how we may use the expectation-maximization (EM) algorithm to form $\widetilde{B}_k$.

Different experts and/or segmentation algorithms will have different segmentation qualities. We represent the quality of each expert, or segmentation, algorithm with a 2×2 matrix λ_k, where $\lambda_k(r,s)$ is the conditional probability that the true classification of the mth pixel is $T(m) = r, m \in \{1,2,\ldots,M\}$, while the kth expert, or algorithm, predicts $B_k(m) = s$. The diagonal entries $(\lambda_k(r,s)$ represent the likelihood of correct classification. The sum $\sum_r \lambda_k(r,r)$ is referred to as the *sensitivity* of the kth algorithm:

$$\text{sensitivity} = \lambda_k(0,0) + \lambda_k(1,1) .$$

The off-diagonal entries $(\lambda_k(r,s), r \neq s)$ represent the likelihood of incorrect classification of the kth expert. The sum $\sum_{r \neq s} \lambda_k(r,s)$ is referred to as the *specificity* of the kth algorithm:

$$\text{specificity} = \lambda_k(1,0) + \lambda_k(0,1) .$$

In the STAPLE algorithm we generate $\widetilde{B}$ and estimate the matrices λ_k, $k \in \{1,2,\ldots,K\}$, using an iterative EM algorithm. In the E-step of the algorithm we estimate the *a posteriori* probability of the ground true segmentation T. Then in the

M-step we estimate the matrices $\lambda_k, k \in \{1,2,\ldots,K\}$, by maximizing the likelihood of the ground truth segmentation which we estimated in the preceding E-step.

Mathematically, the E and M-steps are as follows:

E-step. Let $\lambda_k^{(t)}, k \in \{1,2,\ldots,K\}$, denote the current estimate of the qualities of the individual binary images $B_k, k \in \{1,2,\ldots,K\}$. Then, the *a posteriori* probability that the ground true segmentation $T(m) = r$ is

$$p^{(t)}(T(m) = r | B_1(m), B_2(m), \ldots, B_K(m), \lambda_1^{(t)}, \lambda_2^{(t)}, \ldots, \lambda_K^{(t)}) ,$$

which for simplicity we write as $p(T(m) = r | \{B_k(m)\}, \{\lambda_k^{(t)}\})$. To calculate the *a posteriori* probability we decompose it as follows:

$$p^{(t)}(T(m) = r | \{B_k(m)\}, \{\lambda_k^{(t)}\}) = \frac{\alpha p^{(t)}(T(m) = r | \{\lambda_k^{(t)}\})}{\alpha p^{(t)}(T(m) = r | \{\lambda_k^{(t)}\}) + \beta p^{(t)}(T(m) \neq r | \{\lambda_k^{(t)}\})} ,$$

where

$$\alpha = p^{(t)}(B_1(m) = r, B_2(m) = r, \ldots, B_K(m) = r | T(m) = r, \lambda_1^{(t)}, \lambda_2^{(t)}, \ldots, \lambda_K^{(t)}) ,$$
$$\beta = p^{(t)}(B_1(m) \neq r, B_2(m) \neq r, \ldots, B_K(m) \neq r | T(m) \neq r, \lambda_1^{(t)}, \lambda_2^{(t)}, \ldots, \lambda_K^{(t)}) ,$$

Assuming the experts and/or segmentation experts are independent, then the expressions for α and β become

$$\alpha = \prod_{k:B_k(m)=r} p^{(t)}(B_k(m) = r | T(m) = r) \times \prod_{k:B_k(m)\neq r} (1 - p^{(t)}(B_k(m) = r | T(m) = r)) ,$$
$$\beta = \prod_{k:B_k(m)\neq r} p^{(t)}(B_k(m) \neq r | T(m) \neq r) \times \prod_{k:B_k(m)=r} (1 - p^{(t)}(B_k(m) \neq r | T(m) \neq r)) .$$

We estimate the *a priori* probabilities $P(T(m) = r | \{\lambda_k^{(t)}\})$, which are, in fact, independent of $\{\lambda_k^{(t)}\}$ as follows:

$$P(T(m) = r | \{\lambda_k^{(t)}\}) \equiv P(T(m) = r) = \frac{1}{KM} \sum_k \sum_m \delta(B_k, r) ,$$

where

$$\delta(a,b) = \begin{cases} 1 \text{ if } a = b , \\ 0 \text{ otherwise} . \end{cases}$$

M-step. We update the parameters $\lambda_k^{(t)}, k \in \{1,2,\ldots,K\}$, by maximizing the likelihood of the current ground truth estimate determined in the E-step. Given the previous ground truth estimates $p^{(t)}(T(m) = r|B_k(m), \lambda_k^{(t)})$, the new estimates for the expert parameters are:

$$\lambda_k^{(t+1)}(r,r) = \frac{\sum_{m:B_k(m)=r} p^{(t)}(T(m) = r|\{B_k(m)\}, \{\lambda_k^{(t)}\})}{\sum_m p^{(t)}(T(m) = r|\{B_k(m)\}, \{\lambda_k^{(t)}\})},$$

$$\lambda_k^{(t+1)}(r,s) = \frac{\sum_{m:B_k(m)\neq r}(1 - p^{(t)}(T(m) = r|\{B_k(m)\}, \{\lambda_k^{(t)}\}))}{\sum_m(1 - p^{(t)}(T(m) = r|\{B_k(m)\}, \{\lambda_k^{(t)}\}))},$$

where $s \neq r$. After several iterations the *a posteriori* probabilities $p^{(t)}(T(m) = r|\{B_k(m)\}, \{\lambda_k^{(t)}\})$ and the parameters $\lambda_k^{(t)}, k \in \{1,2,\ldots,K\}$, converge to their final values. The $\widetilde{B}(m)$ are then assigned to the class with the maximum *a posteriori* probability:

$$\widetilde{B}(m) = \begin{cases} 1 \text{ if } p(T(m) = 1|\{B_k\}, \{\lambda_k\}) \geq 0.5, \\ 0 \text{ otherwise}. \end{cases}$$

.

References

1. Dawant, B.M., Hartmann, S.L., Thirion, J.P., Maes, F., Vandermeulen, D., Demaerel, P.: Automatic 3-D segmentation of internal structures of the head in MR images usning a combination of similarity and free-form transformations: part I, methodology and validation on normal subjects. IEEE Trans. Med. Imag. 18, 909–916 (1999)
2. Rohlfing, T., Russakoff, D.B., Muurer Jr., C.R.: Performance-based classifier combination in atlas-based image segmentation using expectation-maximization parameter estimation. IEEE Trans. Med. Imag. 23, 983–994 (2004)
3. Warfield, S.K., Zou, K.H., Wells, W.M.: Simultaneous truth and performance level estimation (STAPLE): An algorithm for the validation of image segmentation. IEEE Trans. Med. Imag. 23, 903–921 (2004)

Chapter 22
Biometric Technologies

Abstract. In this chapter we provide a brief overview of biometric technology and in particular, multi-modal biometric technology.

22.1 Introduction

We define biometric technology to be the automated use of physiological, or behavioral, characteristics to determine, or verify, an individual's identity. We shall follow common practice and use the word biometric to refer to any human physiological, or behavioral, characteristic which possesses the following properties:

Universal. Every person should have the biometric characteristic.
Unique. No two people should be exactly the same in terms of the biometric characteristic.
Permanent. The characteristic should be invariant with time.
Collectable. The characteristic can be measured quantitatively.
Reliable. The characteristic must be safe and operate at a satisfactory performance level.
Acceptable. The characteristic should be non-invasive and socially tolerable.
Non-circumventable. The system should be robust against impostors.

Biometric characteristics include visual images and other human phenomena such as speech, gait, odour and DNA.

Example 22.1. Biometric Recognition Systems [4]. A typical use of biometric recognition systems is in access control. In this case, we have a person claiming an identity and provides a biometric sample to support this claim. The biometric recognition system acting as an authentication system then classifies the person as either a true claimant or as an imposter.

In fact biometric recognition systems can be used in three distinct ways [4].

Closed set identity. Classifying a given biometric sample as belonging to one of K persons, where K is the number of known persons.

Open set identity. Assigning a given biometric sample to one of $K+1$ classes, where the first K classes represent the K persons in the closed set identity and class $K+1$ represents an "unknown" or "previously unseen" person.

Authentication. Assign a given sample into one of two classes: either the sample belongs to a specific person or it does not.

22.2 Multi-modal Biometrics

Although there has been a significant improvement in the performance of single-sensor biometric systems, real-life environments still pose significant difficulties on biometric-based recognition systems.

Example 22.2. Unconstrained Iris Recognition [7, 9]. Iris recognition authenticates and recognizes persons using the unique iris pattern. When the iris image is taken in a controlled environment, the accuracy of the system is very high. However in uncontrolled environments the performance of the system degrades considerably due to noise factors such as eyelids, eyelashes and reflection.

To increase overall reliability, the contemporary biometric systems often employ multiple physiological, or behavioral, traits. This approach is called multi-modal biometrics.

The most often multi-biometric data, employed in the biometric systems, include iris and retina of the eye, fingerprint, geometry and palmprint of the hand, and also face and ears. We often combine complimentary biometric sensors. For example, face geometry is a highly dynamic but rich topological structure (smile, lip, brow, and eye movements). Combining facial images with more static biometric such as fingerprint, is an example of the concept of multi-biometrics.

The multi-modal human recognition is deployed in physical access security systems and other areas such as banking systems [11].

22.2.1 Fingerprints

Fingerprint is, perhaps, the oldest type of biometrics and today the fingerprint reader is the most developed type of biometric sensors.

22.2.2 *Signatures*

Current interest in signature analysis is motivated by the development of improved devices for human-computer interaction which enable input of handwriting and signatures.

22.2.3 *Faces*

Face recognition systems detect patterns, shapes, and shadows in the face, perform feature extraction and recognition of facial identity. Today, the most popular approach is based on principal component analysis (see Chapt. 9).

However many face recognition systems are confused when identifying the same person smiling, aged, with various accessories (moustache, glasses), and/or in badly lit conditions. For a comprehensive review of the current approaches to face recognition see [12].

22.2.4 *Iris and Retina*

Iris recognition systems scan the surface of the iris to compare patterns. The iris biometrics [2] is considered to be the most reliable one. Retina recognition systems scan the surface of the retina and compare nerve patterns, blood vessels and such features [5].

22.2.5 *Gait Biometrics*

Gait recognition is defined as the identification of a person through the pattern produced by walking [3]. A unique advantage of using gait as a biometric is that it offers the potential for recognition at a distance or at low resolution, when other biometrics might not be perceivable. However, this must be balanced with the large within-subject variation exhibited by a gait biometric [3].

22.2.6 *Other Biometrics*

A variety of biometrics such as ear geometry [6], odour, electrocardiogram and keystroke dynamics [11].

22.3 Multi-biometrics

Apart from multi-modal systems, multi-biometrics include multi-sensor, multi-algorithm, multi-instance and multi-sample systems [10, 11].

22.3.1 Multi-sensor System

Multi-sensor systems employ multiple sensors to capture a single biometric trait. The following example illustrates multi-sensor face recognition which is performed by fusing thermal infra-red and visible light face images together.

Example 22.3. Face Recognition by Fusing Thermal Infrared and Visible Imaqery [1]. Thermal infrared (IR) imagery offers a promising alternative to visible, or electro-optical (EO), imagery for face recognition due to its relative insensitivity to changes in illumination. However thermal IR is opaque to glass. This means part of the face may be occluded if the person is wearing eyeglasses.

The IR and EO images are fused together in the wavelet domain (Chapt. 8). In [1] a genetic algorithm is used to find the optimum strategy to combine the information from the two spectra (cf. Ex. 8.4). The objective is to compute a fused image which captures the most salient features from the inputs.

Face recognition is then performed using a conventional principal component analysis of the fused image. The results showed a significant improvement in recognition.

22.3.2 Multi-algorithm System

Multi-algorithm systems use multiple feature extraction and/or matching algorithms on the same biometric data. The idea underlying a multi-algorithm system is that different features and/or matching algorithm emphasize different aspects of the test object. Their combination may therefore give an improved biometric performance. The following example illustrates a multi-algorithm face recognition system which is performed by fusing PCA (principal component analysis) and LDA (linear discriminant analysis) algorithms.

Example 22.4. Decision-level Fusion of PCA and LDA-based Face Recognition Algorithms [8]. Face recognition is often performed using principal component analysis (PCA) (Sect. 9.2) or linear discriminant analysis (LDA) (Sect 9.5). In [8] face recognition is performed by fusing together the outputs of a PCA and a LDA face recognition algorithm.

Let $\mathbf{x}$ denote a given test image written as a one-dimensional vector. We project $\mathbf{x}$ onto the PCA and LDA sub-spaces. Let $\mathbf{u}$ and $\mathbf{v}$ be the corresponding projected vectors. If $\mathbf{U}_k$ and $\mathbf{V}_k$ are, respectively, the kth training pattern in the PCA and LDA sub-spaces, then we compute the corresponding Euclidean distances $d_k^{PCA} = \|\mathbf{u} - \mathbf{U}_k\|$ and $d_k^{LDA} = \|\mathbf{v} - \mathbf{V}_k\|$. The Euclidean distance are normalized and are then combined together to a given fused distance $\widetilde{D}_k$:

$$\widetilde{D}_k = \frac{D_k^{PCA} + D_k^{LDA}}{2} .$$

where

$$D_k^{PCA} = \frac{d_k^{PCA} - \min_l(d_l^{(PCA)})}{\max_l(d_l^{PCA}) - \min_l(d_l^{PCA})} ,$$

$$D_k^{LDA} = \frac{d_k^{LDA} - \min_l(d_l^{(LDA)})}{\max_l(d_l^{LDA}) - \min_l(d_l^{LDA})} .$$

The fused distances $\widetilde{D}_k$ are then used in a nearest neighbor classifier instead of d_k^{PCA} or d_k^{LDA}. The results obtained with $\widetilde{D}_k$ show a significant improvement in performance.

22.3.3 *Multi-instance System*

Multi-instance systems use multiple instances of the same biometric. For example, we may use the fingerprints from the left and right index fingers, or irises from the left and right eye, to verify an individual.

22.3.4 *Multi-sample System*

Multi-sample systems use multiple samples of the same biometric trait obtained with the same sensor. For example a fingerprint system equipped with a small size sensor may acquire multiple dab prints of an individual's finger in order to obtain a complete fingerprint image.

22.4 Epilogue

The aim of the book has been to provide a practical introduction to the theories, techniques and applications of image fusion. In Part I we considered the conceptual theories and ideas which underlie image fusion and in Part II we considered some of the techniques and algorithms which are used in image fusion. In Part III we considered four real-life applications of image fusion. Apart from these four applications, numerous examples of image fusion are given in Parts I and II. The reader is now encouraged to review these examples.

References

1. Bebis, G., Gyaourova, A., Singh, S., Pavlidis, I.: Face recognition by fusing thermal infrared and visible imagery. Image Vis. Comp. 24, 727–742 (2006)
2. Bowyer, K.W., Hollingsworth, K., Flynn, P.S.: Image understanding for iris biometrics: a survey. Comp. Vis. Image Understand. 11, 281–307 (2008)
3. Boulgouris, N.V., Hatzinakos, D., Plataniotis, K.N.: Gait recognition: a challenging signal processing technology for biometric identification. IEEE Sig. Process. Mag. (November 2005)
4. Cardinaux, F., Sanderson, C., Bengio, S.: User authentication via adapted statistical models of face images. IEEE Trans. Sig. Process. 54, 361–373 (2005)
5. Daugman, J.: The importance of being random: statistical principles of iris recognition. Patt. Recogn. 36, 279–291 (2003)
6. Hurley, D.J., Nixon, M.S., Carter, J.N.: Force field feature extraction for ear biometrics. Comp. Vis. Image Understand. 98, 491–512 (2005)
7. A robust eyelash detection based on iris focus assessment. Patt. Recogn. 28, 1630–1639 (2007)
8. Marcialis, G.L., Roli, F.: Decision-level fusion of PCA and LDA-based face recognition algorithms. Int. J. Image Graphics 6, 293–311 (2006)
9. Min, T.-H., Park, R.-H.: Comparison of eyelid and eyelash detection algorithms for performance improvement of iris recognition. In: Proc. Int Conf. Image Proc. (2008)
10. Ross, A.: An introduction to multibiometrics. In: Proc. 15th Eur. Sig. Proc. Conf. (2007)
11. Ross, A., Nandakumar, K., Jain, A.K.: Handbook of multibiometrics. Springer, Heidelberg (2006)
12. Zhao, W., Chelappa, R. (eds.): Face Processing: Advanced Modeling and Methods. Elsevier, Amsterdam (2006)

Index

Printed by Printforce, the Netherlands